PETROLOGY FOR STUDENTS.

London: C. J. CLAY AND SONS,
CAMBRIDGE UNIVERSITY PRESS WAREHOUSE,
AVE MARIA LANE,

AND

H. K. LEWIS,
136, GOWER STREET, W. C.

Glasgow: 263, ARGYLE STREET.
Leipzig: F. A. BROCKHAUS.
New York: MACMILLAN AND CO.

Cambridge Natural Science Manuals.

PETROLOGY FOR STUDENTS:

AN INTRODUCTION TO THE STUDY OF ROCKS UNDER THE MICROSCOPE.

BY

ALFRED HARKER, M.A., F.G.S.,

FELLOW OF ST JOHN'S COLLEGE, AND
DEMONSTRATOR IN GEOLOGY (PETROLOGY) IN
THE UNIVERSITY OF CAMBRIDGE.

CAMBRIDGE:
AT THE UNIVERSITY PRESS.
1895

Cambridge:
PRINTED BY J. & C. F. CLAY,
AT THE UNIVERSITY PRESS.

PREFACE.

THE greatest difficulty that I have experienced as a teacher of petrology has been in recommending a text-book suitable for English students. As an attempt to meet this difficulty the following pages have been written. They are intended as a guide to the study of rocks in thin slices, and are of course assumed to be supplemented throughout by demonstrations on actual specimens. For this reason the examples are chosen, so far as possible, from British rocks.

No systematic account has been given here of the crystallographic and optical properties of minerals: Professor Idding's translation of Rosenbusch's text-book and Dr Hatch's translation of the same author's tables have rendered this unnecessary. In particular I have made no reference to methods depending on the use of convergent light.

I am indebted, as every writer on this subject must be, to the works of Zirkel, Rosenbusch, Fouqué and Lévy, and other authorities, as well as to Mr Teall's "British Petrography"; but, so far as was possible, all descriptions of rocks have been written directly from specimens. Numerous references to original sources have been given in foot-notes, but I have thought it advisable to restrict these references to easily accessible English works.

I have often cited also the coloured plates in some standard works of reference, to which most students will have access. In view of the difficulty of adequately representing rock-sections by means of process-blocks, the figures in this book are selected chiefly to illustrate simple structural characters, and some of them are necessarily rather diagrammatic.

A. H.

ST JOHN'S COLLEGE, CAMBRIDGE.
May, 1895.

CONTENTS.

REFERENCES.

Rosenbusch-Iddings, *Microscopical Physiography of the Rock-forming Minerals* (with photographic plates), 1888.

Fouqué and Lévy, *Minéralogie micrographique* (with atlas of coloured plates), 1879.

Teall, *British Petrography* (with numerous coloured plates), 1888.

Rosenbusch-Hatch, *Petrographical Tables.*

ABBREVIATIONS.

G. M. = Geological Magazine.

M. M. = Mineralogical Magazine.

Q. J. G. S. = Quarterly Journal of the Geological Society.

CHAPTER I.

INTRODUCTION.

In this section will be included such notes on the optical properties of minerals as may be of use to a novice; but there will be no attempt to supersede the use of books dealing systematically with the subject.

Microscope[1]. We shall assume the use of a microscope specially adapted for petrological work, and therefore fitted with polarising and analysing prisms, rotating stage with graduated circle and index, and 'cross-wires' of spider's web properly adjusted in the focus of the eye-piece. The sub-stage mirrors attached to such instruments usually have a flat and a concave face. With day-light the flat face should be used; with artificial light things should be so arranged that the mirror, used with the concave face, gives as nearly parallel rays as possible.

A double nose-piece, to carry two objectives, is very useful, although it usually gives very imperfect centring for high powers. The most useful objectives are a 1 inch or $1\frac{1}{2}$ inch and a $\frac{1}{4}$ inch, but for many purposes a $\frac{1}{8}$ inch is also very desirable. For minute objects, such as the 'crystallites' in glassy rocks and the fluid-pores in crystals, a high power is indispensable, and for very fine-textured sedimentary rocks an immersion-lens offers great advantages.

A selenite-plate, a quartz-wedge, and other special pieces of

[1] For a brief historical sketch of the application of the microscope to petrology see G. H. Williams' pamphlet *Modern Petrography* (Monographs on Education), Boston, 1886.

apparatus will be of use for various purposes. The methods involving their use may be found in the mineralogical text-books; where too the student will find guidance as to the examination of crystal-slices by convergent light.

Form of section of a crystal and cleavage-traces. A well-formed crystal gives in a thin slice a *polygonal section*, the nature of which depends not only upon the forms present on the crystal, but also on the direction of the section and on its position in the crystal, as, *e.g.*, whether it cuts through the centre or only truncates an edge or corner. A cube cut parallel to one pair of faces gives a square, by merely truncating one solid angle of the crystal we get a triangle, and a parallel section through the centre gives a hexagon. Again, the same shape of section may be obtained from very different crystals. Thus we may get a regular hexagon not only from a cube, as remarked, but from an octahedron or a rhombic dodecahedron cut in the same crystallographic direction, or again from a hexagonal prism or pyramid cut perpendicularly to its vertical axis. Nevertheless if several crystals of one mineral are present in a rock-slice, we can by comparison of the several polygonal sections obtain a good idea of the kind of crystal which they represent. Further, if by optical or other means we can determine approximately the crystallographic direction in which a particular crystal is cut, we can usually ascertain what faces are represented by the several sides of the polygon.

For this purpose we may require to measure the angle at which two sides meet, and this is easily done with a microscope provided with a graduated circle. Bring the angle to the intersection of the cross-wires, adjust one of the two sides to coincide with one of the cross-wires, and read the figure at the index of the circle. Then rotate until the other side is brought to coincide with the same cross-wire, and read the new figure. The angle turned through is the angle between the two sides of the section.

This angle is the same as that between the corresponding faces of the crystal only provided the plane of section cuts these two faces perpendicularly. For a section nearly perpendicular to the two faces, however, the error will not be great.

In consequence of the mechanical forces which affect rock-masses, and also as a result of the process of grinding rock-slices, the minerals often become more or less fractured or even shattered. In a strictly homogeneous substance the resulting cracks are irregular, but if there be directions of minimum cohesion in crystals (cleavage), the cracks will tend to follow such directions, and will appear in a thin slice as fine parallel lines representing the *traces of the cleavage-planes* on the plane of section. The regularity and continuity of the cracks give an indication of the degree of perfection of the cleavage-structure, but it must also be borne in mind that a cleavage making only a small angle with the plane of section will, as a rule, not be shown in a slice.

In the case of a mineral like augite or hornblende, with two directions of perfect cleavage, the angle which the two sets of planes make with one another is, of course, a specific character of the mineral, or at least characteristic of a group of minerals, such as the pyroxenes or the amphiboles. In a slice perpendicular to both the cleavages the traces will shew the true angle; for any other direction of section the angle between the cleavage-traces will be different, but it will not vary greatly for slices nearly perpendicular to both the cleavages, and will often suffice for discrimination, as for instance between the 87° of the pyroxenes and the $55\frac{1}{2}$° of the amphiboles. In a slice parallel to the intersection of the two cleavages the two sets of cleavage-traces reduce to one, and a slice of a mineral such as augite or hornblende which exhibits but one set of cleavage-traces may be assumed to be nearly parallel to the 'vertical axis' of the crystal. It may be remarked that minerals having two good cleavages tend to develope in such a way that their greatest elongation is parallel to the intersection of the two cleavages, as in the minerals named. In some cases this is only true of the smaller crystals, *e.g.* in the felspars. Similarly a mineral like the micas with one strongly marked cleavage usually presents a tabular habit with broad faces parallel to the cleavage-direction.

A mineral not possessing any good cleavage often shews irregular cracks in rock-slices (*e.g.* quartz and usually olivine). This is especially the case in brittle minerals.

Sometimes the cracks, though not regular enough to indicate a good cleavage, may have a tendency to follow a particular direction, as in the 'cross-jointing' of apatite and the 'transverse fissures' of tourmaline. Again, a system of 'gliding-planes' may closely imitate the effect of a true cleavage, as in the transverse 'parting' of cyanite, of some augites, *etc.*

Transparency, colours, and refractive indices of minerals. Only a few rock-forming minerals remain opaque even in the thinnest slices: such are graphite, magnetite, pyrites, and pyrrhotite; usually hæmatite, ilmenite, limonite, and kaolin; sometimes chromite or picotite. These should always be examined in reflected light; the lustre and colour, combined with the forms of the sections and sometimes the evidence of cleavage, will usually suffice to identify any of these minerals. The great majority of rock-forming minerals become transparent in thin slices. Those which seen in hand-specimens of rocks appear opaque, are often strongly coloured in slices, while those which in hand-specimens shew colours, are frequently colourless in thin slices. In the case of many minerals these 'absorption-tints' are thoroughly characteristic, but still more so are the differences of colour (pleochroism) in one and the same crystal according to the direction of the slice and the direction of vibration of a polarised beam traversing it, to be noticed below.

The colours ascribed to minerals in the following pages and the epithet colourless apply to thin slices of the minerals.

Apart from colour, the aspect of a mineral as seen in thin slices by natural light varies greatly according to its *refractive index*[1], and it is of great importance for the student to learn to appreciate at a glance the effects due to a high or a low refractive index.

If a thin slice of a single crystal be mounted by itself in some medium of the same colour and refractive index as

[1] By this must be understood its *mean* refractive index. A crystal of any system other than the regular has in any section two refractive indices, the magnitudes of which depend further upon the direction of the section; but these differences in any one mineral are small as compared with the differences between the mean indices in different minerals.

the crystal, its boundaries and surface-characters will be invisible, while its internal structure may be studied to the best advantage. Quartz mounted in Canada balsam (both colourless and of very nearly the same refractive index) is almost invisible. If olivine, a colourless mineral of much higher refractive index, be mounted in balsam, its boundaries and the slight roughness of its polished surface will be very apparent. In ordinary rock-slices, mounted in balsam, a roughened or 'shagreened' appearance may be taken as the mark of a mineral having a refractive index considerably higher than that of the medium used.

Again, a highly refringent mineral surrounded in the slice by others less highly refringent is seen to be more strongly

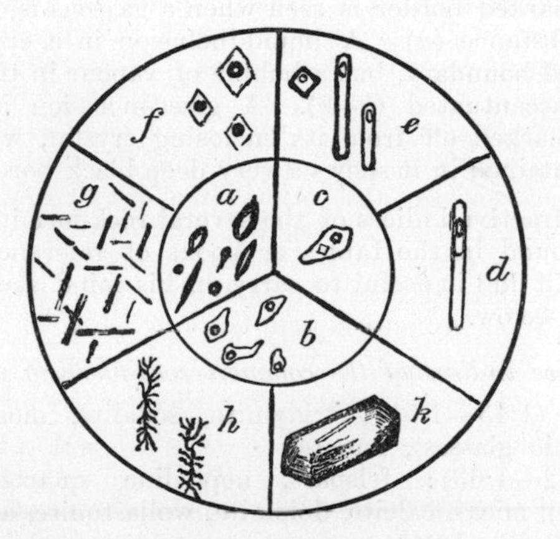

FIG. 1. VARIOUS MICROSCOPIC INCLUSIONS, HIGHLY MAGNIFIED.

a. Gas-pores in obsidian. *b.* Fluid-pores with bubbles; in quartz. *c.* Fluid-pore with bubble and cube of salt; in quartz. *d.* Fluid-cavity in form of negative crystal, containing two fluids and bubble; in quartz. *e.* Fluid-cavities in form of negative crystals, with bubbles; in quartz. *f.* Glass-inclusions in form of negative crystals, with bubbles; in quartz. *g.* Schiller-inclusions consisting of three sets of flat negative crystals filled with opaque iron-oxide; in felspar. *h.* Schiller-inclusions consisting of negative crystals partly occupied by a dendritic growth of iron-oxide; in olivine. *k.* Zircon-crystal enclosed in quartz and itself enclosing an apatite-needle.

illuminated than these, and this brightness is made more conspicuous by a dark boundary which is deeper in proportion to the difference in refractive index between the mineral in question and its surroundings. For these reasons a highly refringent crystal seems to stand out in relief against the rest of the slice (fig. 1 *k*).

Such considerations must be borne in mind in examining the minute inclusions in which many crystals abound. These inclusions may be of gas, of liquid (usually with a gaseous bubble), of glass, or a crystal of some other mineral, and these may be distinguished by observing that the depth of the dark border depends upon the difference in refractive index between the enclosing and the enclosed substance (fig. 1). The most strongly marked border is seen when a gaseous is enclosed by a solid substance (*a*). A liquid-inclusion in a crystal has a less marked boundary, but a bubble of vapour in the liquid is strongly accentuated (*b*—*e*). A glass-inclusion is still less strongly marked off from its enclosing crystal, while a gas-bubble contained in it shews a very deep black border (*f*).

The refractive indices of the several rock-forming minerals may be found in the tables or books of reference, but the student will find it useful to carry in his mind such a list as that given below.

Refractive indices of the common rock-forming minerals.

Very low (1·43—1·51): tridymite, sodalite, most zeolites, (volcanic glasses), leucite.
Low (1·52—1·63): felspars, nepheline, quartz, (Canada balsam), micas, calcite, dolomite, wollastonite, actinolite.
Moderate (1·63—1·645): apatite, tourmaline, andalusite, hornblende.
High (1·68—1·8): olivine, sillimanite, pyroxenes, zoisite, idocrase, epidote, garnets.
Very high (1·9—1·95): sphene, zircon.
Extremely high (2·0—2·7): chromite, rutile.

Extinction between crossed nicols. When the polarising and analysing Nicol's prisms are used together, with their planes of vibration at right angles to one another ('crossed nicols')[1], if no object be interposed, there is total darkness ('extinction'), and the same is the case when a slice of any vitreous substance, such as obsidian, is placed on the stage. If, however, a slice of a crystal of any system other than the regular is interposed, there is in general more or less illumination transmitted, and often bright colours. On rotating the stage[2] carrying the object, it is found that extinction takes place for four positions during a complete rotation, these being at intervals of a right angle. In other words, there are two *axes of extinction* at right angles to one another and the slice remains dark only while these axes are parallel to the planes of vibration of the nicols, which are indicated by the cross-wires in the eye-piece. If we rotate the slice into a position of extinction and then remove the nicols, the cross-wires will mark the axes of extinction in the crystal-slice.

Without attempting to deal fully with this branch of physical optics[3], we may remark that all the optical properties of a crystal are related to three straight lines conceived as drawn within the crystal at right angles to one another (the *axes of optic elasticity*) and to a certain ellipsoid having these three straight lines for axes (the *ellipsoid of optic elasticity*). The positions of the three axes may vary in different minerals, except that they must conform to the symmetry proper to the system, and the same is true of the relative lengths of the axes of the ellipsoid. The plane of section of any slice cuts the ellipsoid in an ellipse, the form and position of which depend upon the direction of the section (*ellipse of optic elasticity*), and the axes of extinction are the axes of this ellipse.

In certain cases the ellipse of optic elasticity may be a

[1] In using the two Nicol's prisms, it should always be ascertained that they are crossed. For this purpose the rotating prisms are usually provided with catches in the proper positions, but the true test is total darkness when no object is interposed.

[2] In some microscopes, such as that devised by Mr A. Dick, the stage is fixed, and the two nicols rotate, retaining their relative position, an arrangement with several advantages. We shall assume for distinctness that the stage is made to rotate, as in the most usual models.

[3] The student is referred for this to such a book as Rosenbusch (transl. Iddings), *Microscopic Physiography of the Rock-making Minerals* (1888), London.

circle. For this any direction is an axis, and accordingly we find that such a slice gives extinction throughout the complete rotation. In crystals of the triclinic, monoclinic, and rhombic systems there are two directions of section which give this result. They are perpendicular respectively to two straight lines in the crystal (the *optic axes*), which lie in the plane of two of the axes of optic elasticity, and are symmetrically disposed towards them. In crystals of the tetragonal and rhombohedral systems the two optic axes coincide with one another and with the unique crystallographic axis, and only slices perpendicular to this give total darkness. In the regular system, the ellipsoid being a sphere, the ellipse is always a circle, and all slices give total darkness between crossed nicols.

Crystals of the regular system are spoken of as singly refracting or optically isotropic, and their optical properties[1] are similar to those of a glassy or colloid substance. Crystals of the other systems are doubly refracting or birefringent, and they are divided into uniaxial or biaxial according as they have one or two optic axes.

It is evident that the chance of a slice cut at random from a birefringent crystal being perpendicular to an optic axis is very small. If more than one crystal of a given mineral in a rock-slice remain perfectly dark between crossed nicols throughout a rotation, it is a safe conclusion that the mineral is a singly refracting one.

Straight and oblique extinction. By bearing in mind that the ellipsoid of optic elasticity, and consequently all the optical properties of a crystal, must conform to the laws of symmetry proper to the crystal-system of the mineral, we can foresee all the important points as regards the position of the axes of extinction in crystals of the different systems cut in various directions. For instance, a longitudinal section of a prism of apatite (a rhombohedral mineral) will extinguish when its length is parallel to either of the cross-wires: this is *straight extinction.* A longitudinal section of a prism of albite (a triclinic mineral) will, on the other hand, have axes of extinction inclined at some angle to its length: this is

[1] That is, such of them as we are here concerned with.

oblique extinction. It is to be noticed that these terms have no meaning unless it is stated or clearly understood from what direction in the crystal the obliquity is reckoned. In these examples we reckoned with reference to one of the crystallographic axes defined by the traces of known crystal-faces. Another character often utilised is the cleavage. Thus in a monoclinic mineral with prismatic cleavages, such as hornblende, we select a crystal so cut that the two cleavages give only one set of parallel traces. These traces are then parallel to one of the crystallographic axes (the vertical axis), and we examine the position of extinction with reference to this. First we bring the cleavage-traces parallel to one of the cross-wires, removing if necessary for this purpose one or both of the nicols, and we note the figure indicated on the graduated circle. Then, with crossed nicols, we rotate until the crystal becomes dark, and again note the figure. The angle through which we have turned is the *extinction-angle*. Observe that if a rotation through, say, 15° in one direction gives extinction, a rotation through 75° in the opposite direction would have given the same. For most purposes we do not need to distinguish between the two directions of rotation, but take merely the smaller of the two angles.

To obtain a measurement of use in identifying a mineral we require more than the above. Slices of a crystal of hornblende cut in various directions along the vertical axis will give different extinction-angles, from zero (straight extinction) in a section parallel to the orthopinacoid to a maximum value in a section parallel to the clinopinacoid. This *maximum extinction-angle* is a character of specific value, being the angle between the vertical crystallographic axis and the nearest axis of optic elasticity. We may determine it with sufficient accuracy for most purposes by noting the extinction-angles in two or three vertical sections of the same mineral in a rock-slice and taking the largest value obtained.

By attention to the following points it is in most cases possible to refer to its crystal-system an unknown mineral of which several sections are presented in a rock-slice.

Regular system : singly refracting; all slices extinguish completely between crossed nicols, as in glassy substances.

Tetragonal and rhombohedral (including hexagonal): birefringent and uniaxial; straight extinction for longitudinal sections of crystals with prismatic habit and for any sections of crystals with tabular habit. The two systems cannot be distinguished from one another by optical tests, but in cross-sections of prisms the crystal outline or cleavages will usually suffice to discriminate.

Rhombic (this and the remaining systems birefringent and biaxial): straight extinction for longitudinal sections of crystals with prismatic habit; sections perpendicular to the vertical axis have axes of extinction parallel to pinacoidal faces or cleavages and bisecting the angles between the traces of prism-faces or prismatic cleavages. A section *nearly* parallel to the vertical axis will give nearly straight extinction, except in minerals (*e.g.* olivine) which have a wide angle between the optic axes.

Monoclinic: two important types may be noticed according as the intersection of the chief cleavages (and direction of elongation of the crystals) lies in or perpendicular to the plane of symmetry. In the former case longitudinal sections may give any extinction-angle from zero up to a maximum value characteristic of the species or variety: in the latter (*e.g.* epidote and wollastonite) longitudinal sections give straight extinction. The former case is the more frequent.

Triclinic: no sections give systematically straight extinction.

Twinning. The existence of twinning in a slice of a crystal is instantly revealed by an examination of the slice between crossed nicols, since the two individuals of the twin shew different interference-tints and extinguish in different positions[1]. When twin-plane and face of association coincide—the most common case—a slice perpendicular to the twin-plane will give in the two individuals of the twin extinction-angles which, reckoned from the line of junction, are equal but in opposite directions. Conversely, a crystal which gives equal but opposite extinction-angles may be assumed to be cut very

[1] The only exceptions are in minerals, like the spinels, optically isotropic, and in cases in which the law of twinning is such that the directions of the axes of optical elasticity are not altered (*e.g.* quartz).

nearly perpendicularly to the twin-plane. If the plane of section cut the twin-plane of a crystal at a very small angle, the two individuals of the twin will overlap for a sensible width, and we shall see between the two a narrow band which does not behave optically with either.

When repeated twinning occurs, as in felspars with albite-lamellation, the lamellæ divide, as regards optical behaviour, into two sets arranged alternately.

Extinction-angles in felspars. The discrimination of the several felspars by means of their extinction-angles measured on cleavage-flakes, as perfected by Schuster, is a method of great precision, but is not applicable to crystals in rock-slices. For these the method advocated by Michel Lévy and others will often be found useful. There are two cases in which it is readily applied.

(i) For crystals with albite-lamellation :—Select sections cut approximately perpendicular to the lamellæ. These are known by the extinction-angles in the two alternating sets of lamellæ, reckoned from the twin-line, being in opposite directions and nearly equal; also by the illumination of the two sets of lamellæ being not very different when the twin-line is parallel to a cross-wire. Measure the angles in question in three or four crystals so selected, and take the greatest value found. This will be very nearly the maximum angle for all such sections, which is a specific constant for each kind of felspar, as indicated below for certain types:

Albite, pure,	Ab	− 16°
Oligoclase of constitution	Ab_4An_1	+ 1°
Oligoclase ,, ,,	Ab_3An_1	+ 5°
Oligoclase-andesine	Ab_5An_3	+ 16°
Labradorite, most acid type	Ab_1An_1	+ 27°
Labradorite, medium	Ab_3An_4	+ 38°
Anorthite, nearly pure	An	+ 53°

The angles corresponding to other members of the albite-anorthite series can be interpolated by means of a curve such as that on p. 13. The signs + and − denote angles measured in opposite directions crystallographically. Unless other means of discrimination can be made use of, we have usually no way of distinguishing the two directions, and there is consequently an ambiguity between albite and the oligoclase-andesines. The other felspars have each a characteristic range of angles; thus:—

0° to 5°	oligoclase, the more acid types,
16° to 22°	andesines,
27° to 45°	labradorites,
45° to 50°	bytownites,
beyond 50°	varieties near anorthite.

It may be remarked further that, unless we discriminate the two directions (at right angles to one another) for which extinction occurs, there is an ambiguity between bytownite and anorthite; for a crystal-lamella which extinguishes at, say, 40° to the right will also extinguish at 50° to the left.

(ii) For microlites, assumed to have their length parallel to the intersection of the two principal cleavages:—Here we measure extinction-angles from the long axis of the microlites, and select the highest angle obtained by measurements on several microlites. The following are the characteristic maxima for certain varieties of plagioclase:—

Albite, pure,	Ab	+ 20°
Oligoclase of constitution	Ab_3An_1	0°
Oligoclase-andesine	Ab_5An_3	− 7°
Labradorite, most acid	Ab_1An_1	− 18°
Labradorite, medium	Ab_3An_4	− 32°
Anorthite, nearly pure,	An	− 64° ?

The values corresponding to other members of the plagioclase series can be roughly interpolated by means of the curve, p. 13. Here the ambiguity arising from positive and negative angles confuses albite with certain andesines and acid labradorites, while that arising from complementary angles confuses the more basic labradorites with the anorthites. The method, however, enables us to recognise at once by their low extinction-angles (0° to 6°) the oligoclases and oligoclase-andesines and by their high angles (beyond 20°) the basic plagioclases.

Zonary banding in felspars. In many rocks the felspars shew between crossed nicols concentric zones roughly parallel to the boundary of the crystal, the successive zones extin-

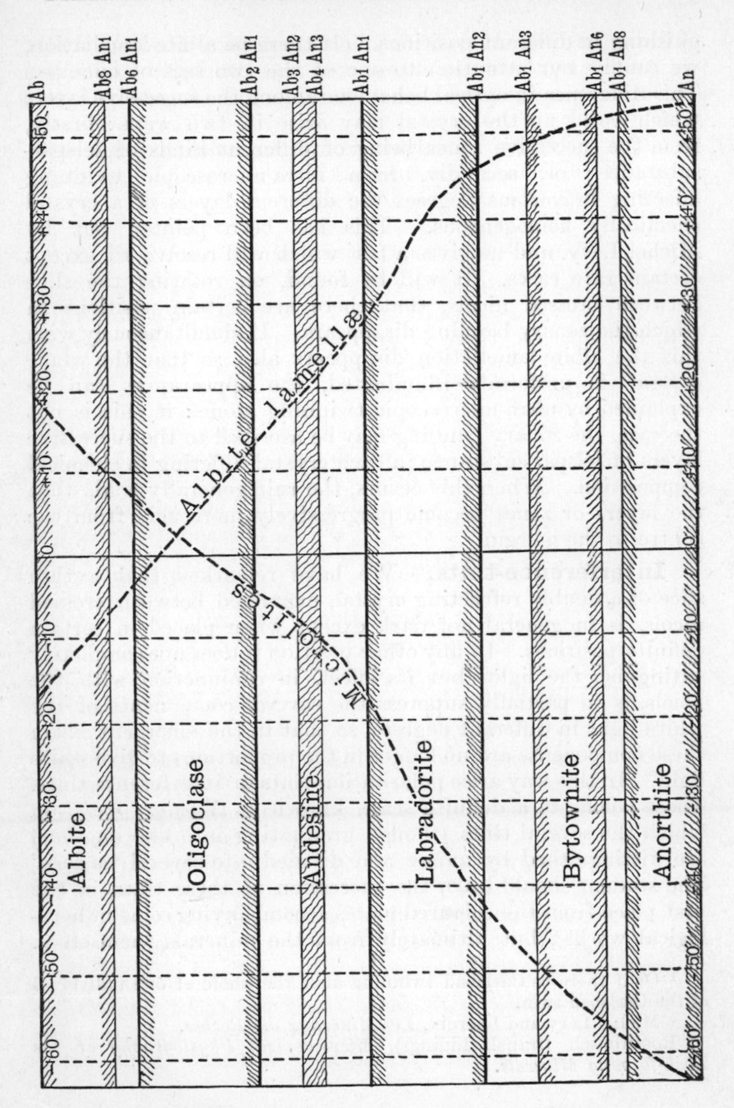

Ab
Ab8 An1
Ab6 An1
Ab2 An1
Ab3 An2
Ab4 An3
Ab1 An1
Ab1 An2
Ab1 An3
Ab1 An6
Ab1 An8
An
Albite
Oligoclase
Andesine
Labradorite
Bytownite
Anorthite
Albite-Lamellae
M on P
+50
+40
+30
+20
+10
0
−10
−20
−30
−40
−50
−60

guishing in different positions. (If there be albite-lamellation, we confine our attention to one of the two sets of lamellæ.) This difference in optical behaviour among the successive layers which build up the crystal may arise in two ways: firstly, from the successive zones being of different kinds of felspar-substance; or, secondly, from ultra-microscopic twinning affecting in various degrees the different layers of a crystal chemically homogeneous. This has been pointed out by Michel Lévy, and he gives a test which will resolve all except certain rare cases. It will be found, on rotating the slice between crossed nicols, that there are certain positions in which the zonary banding disappears. If simultaneously with this the albite-lamellation disappears also, so that the whole crystal[1] is uniformly illuminated, the appearances can be explained by ultra-microscopic twinning alone: if this is not the case, the zonary banding may be ascribed to the successive layers of felspar-substance in each crystal differing in chemical composition. When this occurs, the rule generally holds that the layers or zones become progressively more acid from the centre to the margin.

Interference-tints. We have remarked that a thin slice of a doubly refracting crystal, examined between crossed nicols, is in general not dark except when placed in certain definite positions. In any other position it does not completely extinguish the light, but its effect, in conjunction with the nicols, is to partially suppress the several components of the white light in different degrees, so that in the emergent beam these components are no longer in the proportions to give white light. In this way arise polarisation-tints or interference-tints. These belong to a definite scale, known as *Newton's scale*, on which the several tints (though graduating into one another) are distinguished by names and divided into several 'orders.' The student should learn the succession of these tints, in the first place from the coloured plates accompanying some mineralogical works[2], but ultimately from the minerals themselves.

[1] Or if there be Carlsbad twinning also, the whole of one individual of the Carlsbad twin.

[2] Michel Lévy and Lacroix, *Les Minéraux des Roches*.

Rosenbusch (transl. Iddings), *Microscopical Physiography of the Rock-forming Minerals*.

The precise position in the scale of a given tint observed between crossed nicols can be fixed by means of a quartz-wedge or other contrivance for 'compensating' or neutralising the birefringence of the slice; but for ordinary purposes, at least with colourless or nearly colourless minerals, the interference-tint can be judged by eye with sufficient accuracy. The most brilliant colours are those of the second order and at the top of the first; the lowest colours of the first order are dull greys; while in the third and fourth orders the tints become brighter but paler, ultimately approximating to white.

The interference-tints given by a crystal-section depend (i) on the birefringence of the mineral, which is a specific character; (ii) on the direction of the section relatively to the ellipsoid of optic elasticity, the tint being highest for a section parallel to the greatest and least axes of the ellipsoid; (iii) on the thickness of the slice. These last two are disturbing factors, which must be eliminated before we can use the interference-tints as an index of the birefringence of the crystal, and so as a useful criterion in identifying the mineral.

The fact that the interference-tints depend in part on the direction of the section through the crystal will rarely be found to give rise to any difficulty in roughly estimating the birefringence of the mineral. If two or three crystals of the same mineral are contained in a rock-slice, it is sufficient to have regard to that one which gives the highest interference-tints. Even a single crystal will in the majority of cases give tints not so far below those proper to the mineral as to occasion error, but the possibility of the section having an unlucky direction must be borne in mind.

Rock-slices prepared by a skilful operator are in most cases so nearly constant in thickness that variations in this respect may be left out of consideration. Any important difference is at once detected by well-known minerals giving unusual interference-tints. Thus if quartz or orthoclase give the yellow of the first order, the slice is rather a thick one; if they give orange or red, the slice is considerably thicker than the average of good preparations. Knowing this, we can make allowance for it in estimating the birefringence of some doubtful mineral in the same slice. Such allowance can be roughly judged, or it can be made with considerable precision by means of the large coloured plate of Michel Lévy and Lacroix.

The actual birefringence (numerically expressed) of the several rock-forming minerals, and the interference-tints which they afford in slices of ordinary thickness, are given in numerous books and tables. For rough purposes the student will find it useful to remember about as much as is contained in the following table.

Birefringence and interference-tints of the commoner rock-forming minerals. (The colours given are for slices ·001 inch in thickness.)

Very weak (giving steel-grey tints): leucite, apatite, nepheline.

Weak (giving blue-grey to white of first order): zoisite, microcline, orthoclase, albite, oligoclase, andesine, labradorite, quartz, bytownite, enstatite.

Moderate (giving white, yellow, or orange of first order): andalusite, chlorite, anorthite, hypersthene.

Strong (giving red of first order to violet and blue of second): tourmaline, augite and diallage, common hornblende and actinolite.

Very strong (giving green, yellow, or orange of second order): olivine, epidote, talc, biotite, muscovite.

Extremely strong (giving the pale colours of the third and fourth orders to almost pure white): zircon, hornblende rich in iron, sphene, calcite and dolomite, rutile.

Note that in minerals with strong absorption, such as the deep-coloured micas and hornblendes, the interference-colours are more or less masked by those due to absorption.

Pleochroism. A character often useful in identifying minerals is pleochroism, the property of giving different absorption-tints for different directions of vibration of the light within the crystal. To observe this property, we use the lower nicol only, and rotate either it or the stage. The direction of vibration is that of the shorter diagonal of the nicol.

It is necessary not only to observe the changes of colour, if any, but also to note their relation to directions of vibration within the crystal. For example, elongated sections of biotite and hornblende, tourmaline and sphene, may be found to change from a deeper to a paler tint of brown on rotation; but while in the first pair of minerals the direction of vibration most nearly coincident with the long axis of the section gives the deeper tone, in the second pair it gives the paler.

To be more precise, we wish to know, for a specification of the pleochroism of a given mineral, the absorption-tints for vibrations in three definite directions within the crystal—those of the three 'axes of optical elasticity.' Taking a given mineral, say a hornblende, of which a number of crystals occur in our slice, we may proceed as follows. Select a crystal shewing only one set of cleavage-traces and giving the maximum extinction-angle: this section will be approximately parallel to the plane of symmetry, and will contain two of the required axes. These axes are the axes of extinction for the section, and their positions are thus easily found. The one nearest to the cleavage-traces is the γ-axis, the other the α-axis. Bring the γ-axis to coincide in direction with the shorter diagonal of the nicol, adjusting the position by obtaining extinction, and then removing the upper nicol. Observe the colour: then do the same for the α-axis. For the remaining β-axis we must use another crystal. We may choose one shewing only a single set of cleavage-traces and giving straight extinction: the β-axis is perpendicular to the cleavage-traces. Or we may choose a section shewing two sets of cleavage-traces intersecting at a good angle and extinguishing along the bisectors of the angles between the cleavage-traces: the β-axis is the bisector of the acute angle. The results may be expressed thus in a 'scheme of pleochroism':

α, pale straw,
β, deep brown with greenish tinge,
γ, deeper greenish-brown.

Or we may use the 'absorption-scheme':

$$\gamma \geqq \beta >> \alpha,$$

signifying that the absorption parallel to γ is slightly greater than that parallel to β, and this considerably greater than that parallel to α.

Minerals of the rhombohedral and tetragonal systems can have only two distinct absorption-tints (*dichroism*), one for vibrations parallel to the longitudinal axis (extraordinary ray), the other for vibrations in any direction perpendicular to it (ordinary ray). Thus a particular variety of tourmaline may give

E, colourless,
O, pale indigo;

or absorption-scheme

$$O > E.$$

In minerals of the regular system there can be no pleochroism.

In consequence of pleochroism the absorption-tints of a mineral vary in different crystals when seen in natural light, but the precise nature of the pleochroism can be investigated only with a polarized beam. If the pleochroism is feeble, it is best seen by rotating the nicol, not the stage, but it is not safe to use this method with day-light, which is often partly polarized.

Examination of a rock-slice. In studying a rock-slice it is always well to proceed methodically. A low power should first be used: any object which it is desirable to examine under a higher magnification should be brought to the centre of the field before the objective is changed for a higher power. The slice should always be observed first in natural light: by their outline, relief, cleavages, inclusions, alteration-products, *etc.*, all the ordinary rock-forming minerals can be identified in most cases without the use of polarised light. If the lower nicol is not readily movable, it may be left in for many purposes; but it must be remembered that half the illumination is thus cut off, and for any but the lowest magnifying powers this is of importance. Opaque substances should always be viewed in reflected light.

To examine the pleochroism of any coloured constituent, we put in the lower nicol, and rotate either it or the stage. For verifying feeble pleochroism the former plan is preferable, but the nicol must be rotated until its catch holds it before proceeding to the use of the two nicols, which will be the next act. (Note, however, the caution given above.)

For some purposes oblique illumination is advantageous. For instance, the extremely slender needles of apatite in certain lamprophyres and other rocks become visible only by this means. A 'spot-lens' may be improvised by placing beneath the stage a convex lens of short focal length with its central part covered by a disc of black paper.

In using a high power it will be noticed that the focus is very perceptibly different for the upper and lower surfaces of the slice. To make out the form of a body enclosed in the thickness of the slice the focus should be gradually moved, so as to bring different depths successively into view.

It cannot be too strongly insisted that the identification of the component minerals of a rock is only a part of the examination. The mutual relations of the minerals and their structural peculiarities must also be observed; the order of consolidation, intergrowths, interpositions, decomposition-products, pseudomorphs, *etc.*, as well as special rock-structures such as fluxion-phenomena, vesicles, effects of strain and fracture, *etc.* In short, the object of investigation should be not merely the composition of the rock, but its history.

Classification and nomenclature of rocks. Petrology has not yet arrived at any philosophical classification of rocks. Further, it is easy to see that no classification can be framed which shall possess the definiteness and precision found in some other branches of science. The mathematically exact laws of chemistry and physics which give individuality to mineral species do not help us in dealing with complex mineral aggregates, and any such fundamental principle as that of descent, which underlies classification in the organic world, has yet to be found in petrology. Rocks of different types are often connected by insensible gradations, so that any artificial classification with sharp divisional lines cannot truly correspond to nature. At present, therefore, the best arrangement is that which brings together as far as possible, for convenience of description, rocks which have characters in common, the characters to be first kept in view being those which depend most directly upon important genetic conditions. The grouping adopted below must be regarded as one of convenience rather than of principle.

In a perfect system the nomenclature should correspond with the classification. This is of course impossible at present in petrology. Moreover great confusion has arisen in the nomenclature of rocks in consequence of the rapid growth of descriptive petrography. Many of the names still in use are older than the modern methods of investigation: they were given at a time when trivial distinctions were emphasized, while rocks essentially different were often classed together. Later writers, each in his own way, have arbitrarily extended, restricted, or changed the application of these older names, besides introducing new ones. The newer rock-names need cause no confusion, provided they are employed in a strict sense. Thus 'foyaite' should be used for rocks like that of Foya, specimens of which are in every geological museum: to extend the name to all nepheline-bearing syenites is to introduce needless ambiguity. In practice perhaps the most convenient usage is to speak of 'the Foya type,' 'the Ditro type,' *etc.*, referring in each case to a described and well-known rock. There remain the names employed for families of rocks: some of these are old names, such as granite and syenite, which have come to have a tolerably well understood signification, not always that first attached to them; others, such as peridotite, have been introduced to cover rocks not recognised as distinct families by the earlier geologists. A division of a family is often designated by prefixing the name of some characteristic mineral of that division; *e.g.* hornblende-granite, hypersthene-andesite, *etc.*

These remarks apply more especially to igneous rocks, which we shall consider first. Such rocks, formed by the consolidation of molten 'magmas,' differ from one another in character, the differences depending partly on the composition of the magma in each case, partly on the conditions attending its consolidation. The composition is to some extent indicated by the essential minerals of the rock, which thus become an important, though not logically a prime, factor in any genetic classification. It is evident, however, that a mere enumeration of the minerals of a rock, without taking account of their relative abundance, cannot give a very precise idea of the bulk-analysis[1]; while, on the other hand, it appears on exami-

[1] This difficulty is only partially evaded by ranking some of the constituent minerals as *essential* and others as *accessory*.

nation that magmas of very similar composition may, under different conditions of consolidation, give rise to widely different mineral aggregates. Again, many rocks consist only in part of definite materials, the residue being of unindividualised matter or 'glass.'

To diverse conditions of consolidation must be referred differences in coarseness or fineness of texture, the presence or absence of any glassy residue, the evidence of one or more than one distinct stage in the solidification, and, in general, the peculiarities in the mutual arrangement of the constituent minerals, which collectively are termed the 'structure' of the rock.

We shall first divide the massive igneous rocks into *plutonic*, *intrusive*, and *volcanic*, these names expressing the different geological relations of the rocks as seen in the field, but the divisions themselves being based upon the structural characters which different conditions of consolidation have impressed upon the rocks. Under each of these three heads the various rock-types will be grouped in families founded proximately on the mineralogical, ultimately on the chemical, composition, though this cannot be done without some few inconsistencies. The families will be arranged roughly in order from the more acid to the more basic, but it must be remembered that such an arrangement in linear series can represent only very imperfectly the manifold diversity met with among igneous rocks.

A. PLUTONIC ROCKS.

THE rock-types to be treated under the head of plutonic are met with, in general, in large rock-masses which have evidently consolidated at considerable depths within the earth's crust. Transgressive as regards their actual upper boundary, their geological relations on a large scale are, as a rule, only imperfectly revealed by erosion; so that their actual form and extent are often matters of conjecture. Some of the masses seem to be of the nature of great laccolites; others have been supposed to mark reservoirs of molten magma which once furnished the material of minor intrusions and surface volcanic ejectamenta. The immediate apophyses of the large masses have similar petrological characters.

The distinctive features of these rocks of deep-seated consolidation are those which point to slow cooling (not necessarily slow consolidation) and great pressure. The rocks are without exception *holocrystalline*, *i.e.* they consist wholly of crystallized minerals with no glass. Even as microscopic inclusions in the crystals, glass is much less characteristic than water, which gives evidence of high pressure during the crystallization. The texture of plutonic rocks may be comparatively coarse, *i.e.* the individual crystals of the essential minerals may attain considerable dimensions. The typical structure is that known as *hypidiomorphic*, only a minor proportion of the crystals being 'idiomorphic' (*i.e.* developing their external forms freely), while the majority, owing to mutual interference, are more or less 'allotriomorphic' (taking their shape from their surroundings)[1]. Other structural features will be noticed below.

[1] This is the terminology used by Rosenbusch. Zirkel has adopted Rohrbach's terms *automorphic* and *xenomorphic* in the same senses.

In particular, typical plutonic rocks are *non-porphyritic*, *i.e.* there is evidence of but one continuous stage in the consolidation. In many intrusive and almost all volcanic rocks, some one, or more, constituent (usually a felspar) occurs in two distinct generations with different habits and characters, belonging to an earlier and a later stage of consolidation in which quite different conditions prevailed. This is the 'porphyritic' structure, and is typically wanting among plutonic rocks, which have what has been termed an 'even-grained' character ('körnig' of Rosenbusch).

We shall consider the several families in an order which corresponds roughly to their chemical relationship, beginning with the acid rocks and ending with the ultrabasic.

CHAPTER II.

GRANITES.

THE granites are even-grained holocrystalline rocks composed of one or more alkali-felspars, quartz, and some ferro-magnesian mineral, besides accessory constituents. The rocks are generally of medium to rather coarse grain, and the tendency of the crystals as a whole to interfere with one another's free development gives what Rosenbusch styles the hypidiomorphic structure.

According to their characteristic minerals, after felspars and quartz, the rocks are described as *muscovite-*, *biotite-*, *hornblende-*, and *augite-granites ;* and this division corresponds roughly to different chemical compositions, from more to less acid types. *Tourmaline-granite* must be considered a special modification of the above, and, in particular, of the more acid kinds. With the granites we shall also include certain rocks (*aplite*, *pegmatite*, *greisen*) associated with granites but differing from them in important structural and mineralogical characters, some of them never forming, like the true granites, large bodies of rock.

Constituent minerals. Felspars make up the greater part of a granite, a potash- and a soda-bearing felspar commonly occurring together. The potash-felspar is often *orthoclase*, either in simple crystals or in Carlsbad twins, the Baveno twin being uncommon. When fresh, it shews its cleavages and sometimes a slight zonary banding, but these appearances are lost when the mineral is altered to any extent. The common decomposition-processes give rise either to finely divided kaolin or to minute flakes of mica. When the latter are large enough to be clearly distinguished, they are often seen to lie along the cleavage-planes of the felspar. Decomposition often begins in the interior of a crystal, which may be clouded or completely obscured while the margin remains clear. Instead of orthoclase[1] we often find *microcline*, which is usually the last product of consolidation in the rock. When fresh, microcline shews its characteristic 'cross-hatched' struc-

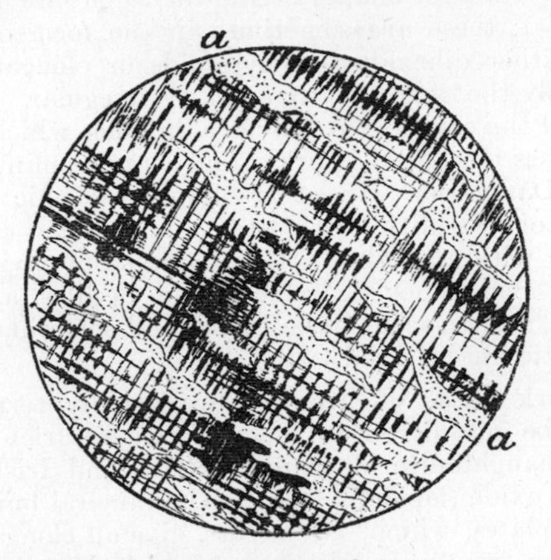

FIG. 2. MICROCLINE FROM THE 'RAPAKIWI' GRANITE OF FINLAND; ×20. crossed nicols: shewing the characteristic 'cross-hatching.' It is traversed by veinlets of albite (*a*) intergrown with crystallographic relation to the microcline [1031].

ture and sometimes a vein-like intergrowth of albite (fig. 2). Its alteration by weathering is similar in kind to that of orthoclase. The soda-felspar of most granites ranges from *albite* to *oligoclase*. It has rather a tabular habit, giving rise to elongated rectangular sections. It is always twinned on the albite- and occasionally too on the pericline-law. The common decomposition-products are kaolin, sometimes paragonite mica, and in the lime-bearing varieties some epidote or

[1] In the British granites in particular microcline is often the prevalent potash-felspar.

calcite. Parallel intergrowths of orthoclase and plagioclase are sometimes found. The felspars of granite are not rich in inclusions, but they may enclose sparingly microlites of the earlier constituents of the rock.

The *quartz* of granites does not usually shew any crystal boundaries, except on the walls of drusy cavities ('miarolitic' structure), or less perfectly when the mineral is moulded by microcline. Its most characteristic inclusions are fluid-cavities (fig. 1, *b*—*e*); these are sometimes in the form of 'negative crystals,' either dihexahedral pyramids or elongated prisms; more usually the shape is rounded or irregular. The liquid does not fill the cavity, but leaves a bubble, which is mobile. In some cases the liquid is brine, and contains minute cubes of rock-salt (Dartmoor). In others liquid carbonic acid occurs instead of, or in addition to, water, and in some cases we see one bubble within another. Glass- and stone-cavities are less abundant. Sometimes extremely fine needles are enclosed (Peterhead): these seem to be rutile, and sometimes shew the characteristic knee-shaped twin.

The dark micas of granites are usually termed *biotite*. This may be considered to include varieties rich in ferrous oxide (the haughtonite of many Scottish and Irish granites), or in ferric oxide (lepidomelane). The mineral builds roughly hexagonal plates, which, cut across, give an elongated section shewing the strong basal cleavage. A lamellar twinning parallel to the base is probably common, but, owing to the nearly straight extinction, this is not often conspicuous. The fresh biotite is deep brown with intense pleochroism. Its common inclusions are apatite, zircon, and magnetite, and the minute zircons are always surrounded by a 'halo' of extremely deep colour and intense pleochroism (Skiddaw, Dartmoor, Dublin, *etc.*). Decomposition often produces a green coloration and ultimately a green chloritic pseudomorph with secondary magnetite-dust. This magnetite may be reabsorbed, restoring the brown colour but with less pleochroism and with loss of cleavage.

The colourless, brilliantly-polarising *muscovite* forms rather ragged flakes, posterior to the biotite or partly in parallel intergrowth with it (Dublin, *etc.* fig. 3, *B*). It is always clear, and is not susceptible to weathering. A lithia-mica, in large flakes, takes the place of muscovite in some greisens and pegmatites.

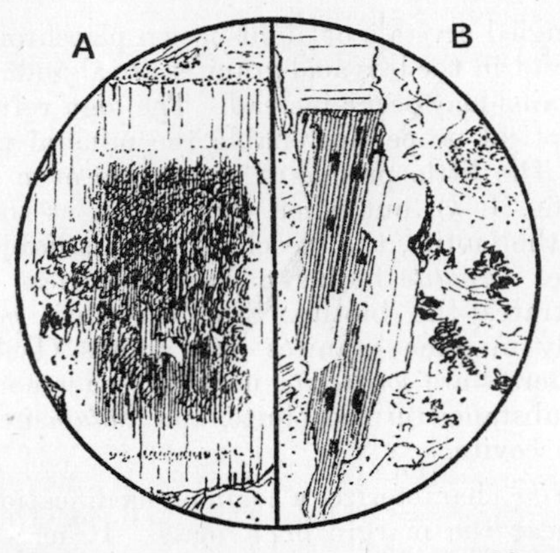

FIG. 3. GRANITE, NEAR DUBLIN; ×20.

A. Crystal of oligoclase shewing zonary structure and decomposition beginning in the interior [389]. *B.* Parallel intergrowth of biotite and muscovite [1774].

The crystals of *hornblende* are irregularly bounded, or at least without terminal planes. They shew the prismatic cleavage, and occasionally lamellar twinning parallel to the orthopinacoid. The colour is green or brownish-green, with marked pleochroism, and the extinction-angle in longitudinal sections always low. Besides inclusions of earlier minerals, there may be an intergrowth with biotite. The common decomposition-products are a green chloritic substance or an epidote and quartz.

When *augite* occurs, it is commonly the variety malacolite or salite, colourless in slices. It is not usually in perfect crystals, but an idiomorphic green augite is found in some coarsely granophyric types of rock (Mull). Augite may be either uralitized or decomposed into a green chloritoid product or serpentine and calcite. The mineral is sometimes accompanied by *enstatite* (Cheviot).

Iron-ores are not plentiful in granites. *Magnetite* may occur or *hæmatite*, either opaque or deep-red; *pyrites* is also found as an original mineral.

Acute-angled crystals of light-brown pleochroic *sphene* are often seen, and in the less acid granites are abundant (fig. 5, *B*). Rounded grains may occur instead. The high refractive index and other optical properties enable the mineral to be readily identified. The little prisms of *zircon* are even more highly refractive (fig. 1, *k*), but when they occur, as they often do, enclosed in the biotite, the pleochroic halo is liable to obscure their nature. *Apatite* builds narrow colourless prisms, and often penetrates the biotite. Small reddish *garnets* occur exceptionally in some granites and aplites (Dublin): other unusual minerals are *cordierite*, usually pseudomorphed by the micaceous substance termed pinite, and *andalusite*, coated with flakes of muscovite.

Tourmaline characterizes a common modification of granite, especially near the margin of a mass. It may be in good crystals but has more frequently ragged outlines. The rude cross-fracture is often apparent. The colour is brown, sometimes with patches of blue, and the dichroism is strong, the strongest absorption being for vibrations transverse to the long axis (the 'ordinary' ray).

Structure. There is in plutonic rocks a normal order of consolidation for the several constituents, which holds good with a high degree of generality. It is in the main, as pointed out by Rosenbusch, a law of 'decreasing basicity.' The order is briefly as follows.

I. Minor accessories (apatite, zircon, sphene, garnet, *etc.*) and iron-ores.

II. Ferro-magnesian minerals:—olivine, rhombic pyroxenes, augite, ægirine, hornblende, biotite, muscovite.

III. Felspathic minerals:—plagioclase felspars (in order from anorthite to albite), orthoclase (and anorthoclase).

IV. Quartz, and finally microcline.

In most rocks such minerals as are present follow the above order. The most important exceptions are the intergrowth of orthoclase and quartz and the crystallization of quartz in advance of orthoclase in some acid rocks, and the rather variable relations between groups II. and III. in some more basic rocks. The order laid down applies in general to parallel intergrowths of allied minerals: thus when augite is intergrown with ægirine or hornblende, the former mineral forms the kernel of the complex crystal and the latter the outer shell; when a plagioclase crystal consists of successive layers of different compositions, the layers become progressively more acid from the centre to the margin.

Certain constituents having variable relations are omitted from the foregoing list. Thus nepheline (elæolite) and sodalite belong to group III., but may crystallize out either before or after the felspars.

The order of consolidation is an important factor in the structure of a rock, *i.e.* in the mutual relations of its constituent parts. In the granites the *normal order* rules with few exceptions. The minor accessory minerals crystallized out first, and are thoroughly idiomorphic, *i.e.* have taken their

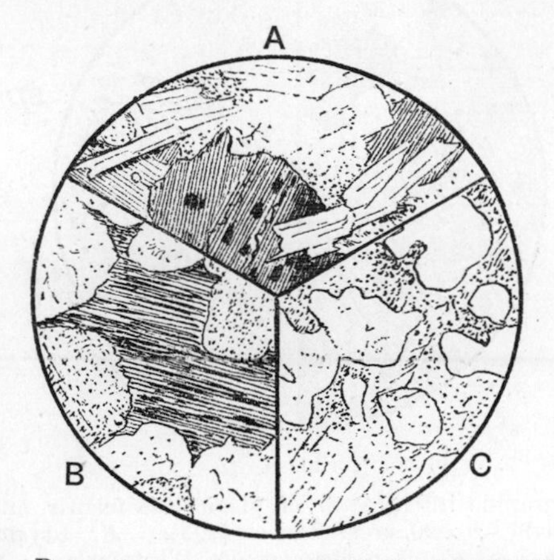

FIG. 4. REVERSALS OF NORMAL ORDER OF CONSOLIDATION IN GRANITES; ×20.

A. Biotite moulded on muscovite, Rubislaw, Aberdeen [390 *a*]. *B.* Biotite moulded on quartz and felspars, Meillionydd, near Sarn, Caernarvonshire [814]. *C.* Orthoclase moulded on quartz, Shap [892].

shape without external interference. The ferro-magnesian minerals preceded the felspars, being often embraced or even enclosed by them, although the felspars may tend also to take crystal outlines. A late crystallization of biotite, *e.g.*, is a rare occurrence (fig. 4, *A*, *B*). Quartz follows the felspars in general. In many granitic rocks quartz-grains may be seen moulded by a felspar (fig. 4, *C*), but it is often found that the latter mineral is microcline.

In other cases the quartz and orthoclase shew an intricate interpenetration owing to their having crystallized simultaneously. Within a certain area of a slice the quartz of such an intergrowth behaves optically as if it were a single crystal, the whole becoming dark between crossed nicols in one position. On rotation the felspar can be made to extinguish in its turn. This is the *micrographic* or micropegmatitic structure, reproducing on a small scale the peculiar arrangement of 'graphic

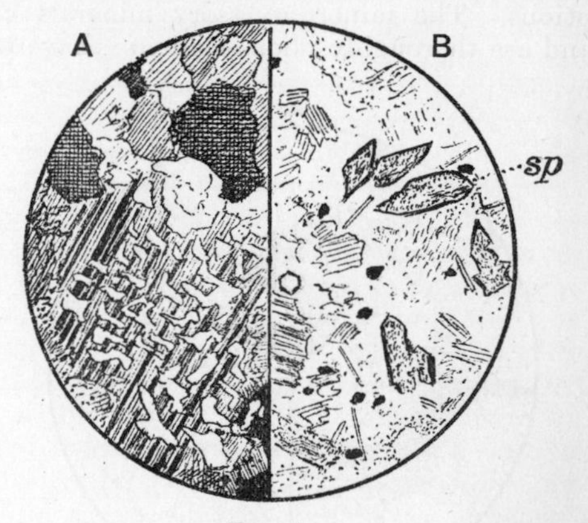

FIG. 5.

A. Micrographic intergrowth of plagioclase felspar and quartz in granite, St David's; × 20, crossed nicols [293]. *B*. Crystals of sphene (*sp*) in dark basic secretion in Shap granite, Westmorland; × 20 [1070].

granite' (pegmatite of Haüy). It is termed by Rosenbusch the granophyric structure. A similar intergrowth of quartz and plagioclase is also found (fig. 5, *A*).

It may be remarked of holocrystalline igneous rocks in general that, when the later crystallized minerals have had a tendency to idiomorphism, vacant spaces are apt to occur, into which project the sharp angles of well-formed crystals. This *miarolitic* or drusy structure is more or less marked in some granites (*e.g.* the Mourne Mts in Ireland).

Some granites have a *porphyritic* character, due to the early development of large crystals of orthoclase (*e.g.* Dartmoor and Shap).

As a more exceptional structure in certain granites may be mentioned the *spheroidal* or orbicular, as in the hornblende-granite of Mullaghderg, Donegal[1].

The parallel and banded structures of some granitic rocks (gneissic granites) will be referred to in a later section.

Leading types. We are able to illustrate all the chief features of granitic rocks by British examples. Almost all the true granites contain a brown mica. If a white mica be present in addition, we have *muscovite-granite* ('two-mica granite' or 'granite proper' of the Germans, 'granulite' of the French[2]). Such rocks are commonly somewhat more acid in composition than those with dark mica only. The Carboniferous granites of Cornwall and Dartmoor afford good examples. They consist of orthoclase, a plagioclase, quartz, and two micas, with the normal order of consolidation. The quartz has fluid-cavities often enclosing minute cubes of rock-salt[3] (Dartmoor, fig. 1, *c*). Parallel intergrowths of biotite and muscovite are common. The minor constituents of the rock are magnetite, apatite, and zircon, the last when it is enclosed in the biotite being always encircled by the characteristic halo of intense pleochroism. More exceptional accessory minerals are andalusite, in pleochroic crystals coated by flakes of muscovite (Cheesewring), and 'pinite' pseudomorphs after cordierite (Land's End). Tourmaline is common, and the rocks graduate

[1] Hatch, *Q. J. G. S.* (1888) xliv., 548—559, Pl. xiv., with a summary of information on other spheroidal granites.

[2] The granulite of German and English writers has a different signification.

[3] Hunt, *G. M.* 1894, 102—104, with figures.

into tourmaline-granites, especially near the margin of an intrusion.

The post-Ordovician granites which occupy so large a tract in Leinster[1] (*e.g.* Dalkey near Dublin) are of a different type. They also have two micas, often in parallel intergrowth, and apatite and zircon are characteristic accessories; but the potash-felspar is microcline, and is the latest product of consolidation. A plagioclase felspar is plentiful, and exceptionally albite is the only felspathic element present (Croghan Kinshela in Wexford). Little crystals of garnet occur in some instances (Three Rock Mountain near Dublin). This mineral is found also in the granite of Foxdale in the Isle of Man, a closely similar rock, in which the dark mica is very subordinate to the white. Another well-known microcline-bearing rock is the 'grey Aberdeen granite' of Rubislaw, *etc.* Similar rocks are found in Donegal.

Rocks in which muscovite is only sparingly or occasionally present form a link with the next division. The Skiddaw granite is of this character. Here the quartz is in great part of prior consolidation to the orthoclase, or there may be some micrographic intergrowth of the two minerals. Felspar-quartz-rocks free from mica are found among the pre-Cambrian intrusions of Ercal in the Wrekin district and of the Malverns. Here too the quartz has crystallized, or has finished crystallizing, before the dominant felspar, which is often microcline. These rocks seem to have affinities with the pegmatites.

It is not necessary to go abroad for examples of muscovite-granites, but we may mention the so-called 'protogine' of M. Blanc, *etc.*, in which the white mica has been erroneously identified as talc.

The commonest division of the granite family is perhaps *biotite-granite* (Fr. granite, Ger. Granitit), characterized by containing a brown mica to the exclusion of muscovite, hornblende, or augite. Such a rock may consist, *e.g.* of orthoclase, albite or oligoclase, quartz, biotite, and minor accessories, with the normal order of consolidation.

The relative proportions of the several minerals vary

[1] Sollas, *Trans. Roy. Ir. Acad.* (1891) xxix., 427—512.

considerably. In the granites (Ordovician and perhaps some older) of Wales quartz is very abundant, and biotite (often destroyed) is only sparingly found. The dominant felspar is often a plagioclase (Caernarvon, St. David's, *etc.*), and probably some of these rocks would be placed among the 'soda-granites' of certain authors.

Departures from the normal order of consolidation are not frequent. Sometimes the quartz has crystallized before the orthoclase, as in the Shap[1] and Eskdale granites, the former of which is further noteworthy for its abundant sphene. Sometimes there is a tendency to a micrographic intergrowth of felspar and quartz (Eskdale granite of Cumberland and St David's).

In many of our British biotite-granites microcline partly or wholly replaces orthoclase (Peterhead, Ross of Mull, Malvern, *etc.*). The Eskdale rock, of Old Red Sandstone age, has little albite veins intergrown in microcline or in orthoclase (microperthite). This feature is better seen in the microcline of some coarse-grained foreign granites, such as the rock styled 'Rapakiwi' in Finland (fig. 2). The probably Tertiary biotite-granite of the Mourne Mountains[2], often micropegmatitic, is specially characterised by its miarolitic structure, with well-formed crystals in its druses.

Less abundant than the types characterised by micas, and usually of less acid composition, is *hornblende-granite* (Ger. Amphibolgranit), in which the distinctive mineral is a green hornblende, usually with biotite in addition. Some of the newer Palæozoic granites of Scotland are of this kind, such as that of Lairg in Sutherland and the Criffel rock at Dalbeattie, in which, however, biotite is predominant. The rock quarried at Mount Sorrel in Charnwood Forest, Leicestershire, is also in part a hornblende-granite, having that mineral associated with biotite. In Ireland a hornblende-granite has been described from Donegal[3], and another is associated with

[1] Teall, Pl. xxxv., fig. 1 [395].

[2] The neighbouring granite of the Carlingford district (Barnavave) is closely similar; Sollas, *Trans. Roy. Ir. Acad.* (1894) xxx. 490.

[3] Hatch, *Q. J. G. S.* (1888) xliv., 548—551.

the Palæozoic biotite-granites of Newry (at Goragh Wood). Hornblende-granites of Tertiary age are found in Skye. Among foreign rocks the red granite of Assouan on the Nile is well known as the material of the Egyptian monoliths, *etc.* It consists of microcline, oligoclase, quartz, and green hornblende, with subordinate biotite, and a little sphene, magnetite, and apatite.

The *augite-granites* are a less common type. An example, of Old Red Sandstone age, occurs in the Cheviots[1]. This consists of orthoclase, plagioclase, quartz, augite, biotite, iron-ores, and apatite, the quartz and orthoclase sometimes shewing a micrographic intergrowth. The colourless augite tends to pass over into a pale green, fibrous hornblende, and chloritic decomposition is also found. Some Tertiary rocks in Mull and in the Carlingford district in Ireland have a green augite without biotite, but these rocks tend to have the structure of granophyres rather than of true granites.

Granites in which a rhombic pyroxene is the dominant ferro-magnesian mineral do not seem to have been recorded. Mr Kynaston, however, finds a pale, faintly pleochroic enstatite as a frequent associate of augite in the Cheviot granites.

Special modifications, etc. A feature in many granites is the occurrence of dark, fine-grained, ovoid patches, which are to be distinguished from enclosed fragments, and may be regarded as *basic secretions* of the granite itself[2]. They are rich in the earliest products of consolidation—apatite, magnetite, sphene—and in the ferro-magnesian constituent of the rock[3]. Sometimes, as in the Criffel granite, hornblende is more plentiful as compared with biotite than in the normal rock, and similarly plagioclase felspar is more abundant relatively to orthoclase. The various points may be studied in the granites of Peterhead and Shap (fig. 5, *B*), where these basic secretions are plentiful. The porphyritic felspars of the

[1] Teall, *G. M.* 1885, 112—116; *Brit. Petr.* Pl. XXXIX, fig. 2.

[2] For the dark patches in the Newry granite Prof. Sollas has suggested a different explanation: *Trans. Roy. Ir. Acad.* (1894) xxx, 502—504. He regards them as highly metamorphosed foreign fragments ('xenoliths').

[3] J. A. Phillips, *Q. J. G. S.* (1880) xxxvi, 1—21; (1882) xxxviii, 216, 217.

Shap granite occur also in the dark patches, but they are partially rounded and corroded, the margin of each crystal being replaced by plagioclase and quartz[1].

In contrast with the dark patches may often be observed light-coloured, coarse-grained streaks and veins traversing granitic rocks. These may be regarded as *acid excretions* of the granite. They are poor in the earlier products of consolidation and the ferro-magnesian constituents, and correspondingly rich in the more acid minerals. They may occur in the same rocks which contain basic secretions, and the contrast between the two modifications is then very evident (Criffel). The mutual relations of the quartz and felspar approach those of the rocks called pegmatites, a rude graphic structure being frequently noticeable.

The *pegmatites* proper form vein-like streaks in granitic or gneissic masses, or may expand into considerable bodies of rock. They consist essentially of orthoclase or microcline[2] and quartz, often with white mica and sometimes red garnet. They frequently have an extremely coarse texture, and structurally they are often characterised by an intergrowth of felspar and quartz, which sometimes assumes the regularity of 'graphic granite', the original pegmatite of Haüy. Pegmatites are largely developed in the region of the Archæan gneiss of Sutherland, and again in Forfarshire[3] and other parts of the Highlands.

A rock of similar mineralogical composition but different structure is *aplite* (also named granitel and semi-granite). This may occur in veins in granite, but cutting it and traversing adjacent rocks. It is a fine-textured rock of thoroughly acid composition. The aplites at Dalkey near Dublin consist of microcline and some oligoclase, quartz, muscovite, and red garnet. An aplite at Meldon in Devonshire is of similar character, but instead of garnet contains topaz and some tourmaline.

[1] *Q. J. G. S.* (1891) xlvii, 282.

[2] Fouqué and Michel Lévy, Pl. VII, fig. 1; Rosenbusch-Iddings, Pl. XXV, fig. 1.

[3] Barrow, *G. M.* 1892, 64.

The Crosby dyke[1] in the Isle of Man may be referred here. It consists essentially of a granular mosaic of clear felspars, quartz, and white mica, the dominant felspar being an albite. Besides the abundant small flakes of white mica, some larger hexagonal plates occur, and sometimes scattered quartz-grains or larger felspars. There are also a few garnets of very irregular shapes, giving a sponge-like appearance in section.

The *tourmaline-granites* appear as modifications of more normal granitic rocks[2]. The tourmaline seems to take the place of the mica. As a further modification, the felspars may partly or wholly be replaced by tourmaline and quartz, the former sometimes occurring in little needles with radiate grouping imbedded in clear quartz. The extreme modification is a tourmaline-quartz-rock or *schorl-rock*, in which felspar is wholly wanting, while tourmaline may occur in two or more habits, as crystals or grains and groups of needles. All these types are illustrated in the marginal parts of the Cornish granites. A curious variety known as luxullianite has been described by Prof. Bonney[3]. Here the conversion of felspars into clear quartz crowded with radiate groups of tourmaline needles can be traced in various stages, the little needles, about ·03 inch in length giving pale brown and light indigo colours for longitudinal and transverse vibrations respectively, while a brown tourmaline in distinct grains has been supposed to represent the mica of the granite. A rock from Trowlesworthy Tor shews a similar replacement of felspar, and has in addition irregular patches of isotropic fluor also enclosing needles of tourmaline[4] (fig. 6, *A*).

The rock known as *greisen* (hyalomicte of French writers) consists essentially of quartz and white mica, which seems to be often a lithia-bearing variety. The Cornish greisens are apparently a modification of the granite in the same sense as the tourmaline-rocks are, but with a different result. The

[1] Hobson, *Q. J. G. S.* (1891) xlvii, 440.

[2] For coloured figure of a tourmaline-granite see Fouqué and Michel Lévy, *Min. Micr.* Pl. VIII, fig. 1.

[3] *M. M.* (1877) i, 215—222.

[4] Worth, *Trans. Roy. Geol. Soc. Cornw.* (1884) x, 177—188.

place of the felspar is partly taken by topaz, and tourmaline may also occur. Greisen is also found in connection with the granite of the Scilly Isles. In Grainsgill, Cumberland, it occurs with rather different relations, though still in association with granite[1]. The white mica builds sometimes rather large

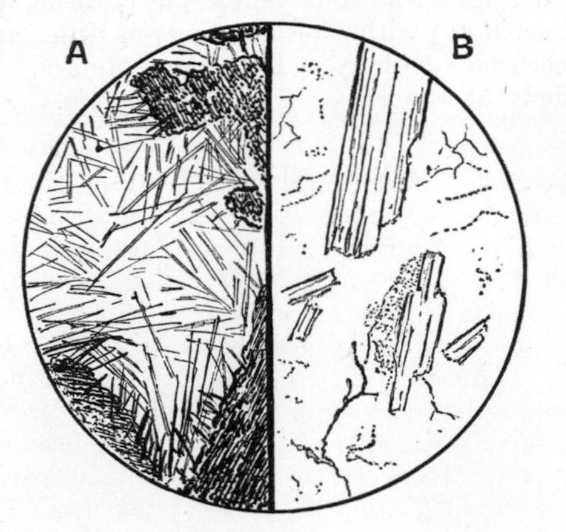

FIG. 6. MODIFICATIONS OF GRANITE; ×20.

A. Replacement of felspar by clear quartz full of tourmaline-needles, Trowlesworthy Tor, Cornwall: with remains of much-decomposed felspar [1361]. *B*. Greisen, Grainsgill, Cumberland: consisting of quartz and muscovite with only occasional relics of turbid felspar [1547].

flakes (fig. 6, *B*), sometimes aggregates of small scales, and in both cases is moulded or enclosed by a moderately coarse mosaic of clear quartz. Tourmaline is absent. A greisen very similar to this is associated with the Foxdale granite in the Isle of Man, and occurs there in the same manner as pegmatite, which is also developed.

The weathering-products of the chief constituent minerals of granite have already been alluded to. The complete destruction of the rock gives rise to a product in which the derivatives of the felspars are the characteristic ingredients. In this way china-clay or kaolin results from the subaerial

[1] Harker, *Q. J. G. S.* (1895), li.

destruction of pegmatite or granite. Another line of alteration is characterised by the production of minerals of the epidote group. The thulite-rock of Trondjhem in Norway is a good example of this. Some felspar and clear quartz remain, but the bulk of the rock consists of thulite (manganese-zoisite) with bright rose and yellow pleochroism, piedmontite (manganese-epidote) with violet and citron tints, and common yellow pleochroic epidote. These, as Reusch[1] shews, are formed chiefly at the expense of the felspars.

[1] *M. M.* (1892) x, 40 (*Abstr.*).

CHAPTER III.

SYENITES (*including* NEPHELINE-SYENITES).

THE syenites are even-grained, holocrystalline rocks consisting essentially of alkali-felspars, and in one group felspathoid minerals, with a smaller proportion of ferro-magnesian constituents and various minor accessories. The texture is often rather coarse to medium-grained, and the structure is that characteristic of plutonic rocks, the several minerals following the normal order of consolidation, and most of them having only imperfect crystal outlines (hypidiomorphic structure of Rosenbusch). In many syenites, however, the order of consolidation is modified by simultaneous intergrowths of different minerals.

This family of rocks is less widely distributed and less abundant than the granites. Considered from a chemical point of view, it is characterised by an unusually high percentage of alkalies. In the syenites which depart farthest in this respect from the common types of igneous rocks, the character shews itself in the presence of felspathoid constituents and soda-bearing ferro-magnesian minerals.

The type characterised by hornblende and alkali-felspars is known as 'syenite proper'[1], or, for clearness, *hornblende-syenite.* When biotite more or less completely takes the place of hornblende, we have *mica-syenite;* and when augite occurs prominently, often in company with one or both of the other coloured minerals, *augite-syenite.* The group characterised by

[1] The original syenite of Werner was the hornblende-granite of Syene or Assouan on the Nile. The name, however, has come to be universally applied to the family under notice, rocks often hornblendic but typically free from quartz.

the occurrence of nepheline or sodalite in addition to felspar is named *nepheline-syenite*, or often elæolite-syenite, without distinction according to the dominant ferro-magnesian constituent, and several types, mostly of restricted occurrence, have received special names. A *leucite-syenite* is known only in the form of rocks with pseudomorphs of orthoclase, elæolite, muscovite, *etc.*, in the form of leucite. The occurrence of subordinate quartz in some syenites gives rise to the varieties *quartz-syenite*, *quartz-mica-syenite*, and *quartz-augite-syenite*, but free silica never occurs in the nepheline-bearing group.

Constituent Minerals. In mode of occurrence, inclusions, alteration-products, *etc.*, the felspars of syenites resemble those of granites. Besides *orthoclase*, *microcline*, and *albite* or *oligoclase*, there occur, especially in the augite- and nepheline-syenites, felspars rich in both potash and soda, known as soda-orthoclase, soda-microcline, anorthoclase, *etc.* These are regarded by some mineralogists as intergrowths on an ultra-microscopic scale of a potash- and a soda-felspar (*cryptoperthite*). An evident parallel intergrowth of albite and microcline or albite and orthoclase (*microperthite*) is also frequent in the same rocks.

When nepheline occurs, it is of the variety known as *elæolite*, in larger and less perfect crystals than the nepheline of volcanic rocks. If idiomorphic, it forms hexagonal prisms with the basal plane bevelled by narrow pyramid-faces. In more shapeless crystals the straight extinction can be verified by reference to rows of inclusions which follow the direction of the vertical axis, and seem to determine the alteration of the mineral. The elæolite is colourless or often rather turbid. It gives rise by decomposition to various soda-zeolites or to moderately brightly polarizing prisms, fibres, and aggregates of cancrinite. A frequent associate of elæolite is *sodalite*, in dodecahedra or in allotriomorphic crystal-plates and wedges. It is colourless or faint blue in slices, and is easily recognised by its isotropic behaviour. It encloses fluid-pores, microlites of ægirine, *etc.*, and secondary products similar to those of elæolite.

The common *hornblende* of syenites is partly idiomorphic but without terminal planes. It is of the green pleochroic variety, giving in vertical sections a maximum extinction-angle of 12° to 16°. Its inclusions and alteration-products are the same as in granite. Some augite-syenites contain the soda-amphibole *barkevicite* with intense brown absorption and pleochroism and an extinction-angle of about 12°.

The *augite*, when it occurs as an accessory, is a very pale green mineral with the same properties as in granite. In the augite-syenites it is sometimes pale green with faint pleochroism, sometimes pale brown to violet-brown with very distinct pleochroism. Various types of schiller- and diallage-structures are sometimes seen, and may affect only a portion—usually the interior—of a crystal. A green pleochroic *ægirine* occurs in some augite-syenites and many nepheline-syenites, and intergrowths of this with augite are not uncommon.

The *biotite* of the syenites is deep brown, becoming green only by secondary changes. In some augite- and nepheline-syenites vibrations parallel to the cleavage-traces are almost completely absorbed. The mineral is roughly idiomorphic, except when intergrown with hornblende or augite.

When *quartz* occurs, it has the same characters as in granite, but is never very abundant. It does not occur in the nepheline-syenites and their allies. Most syenites contain plenty of *sphene* in good crystals shewing the cleavages and often the characteristic twinning[1]. *Zircon* is common in small prisms with pyramidal terminations, as in the granites. In some of the augite-syenites, however, it builds large crystals with good pyramidal shape. It is easily identified by its limpid appearance and extremely high refringence and birefringence. *Apatite* in colourless needles is widely distributed in syenites. The iron-ores are variable in quantity: they include *magnetite*, *ilmenite*, and *hæmatite*, the last two often in thin flakes enclosed in the felspars. An occasional accessory is *perofskite* in small octahedra, distinguished by their very high refractive index and feeble double refraction. Special types contain *melanite* garnet, brown in slices and always isotropic.

Structure. The texture of the syenites and the mutual relations of their constituent minerals are normally similar to those observed in the granites, Rosenbusch's 'order of

[1] See Rosenbusch-Iddings, Pl. XI, fig. 3; XXIII, fig. 1.

consolidation' being, as a rule, followed. In the typical hornblende-syenites there are few peculiarities. When quartz enters, it may be intergrown in micrographic fashion with part of the orthoclase, and this is specially the case in some augite-syenites. When plagioclase felspar is abundant, it is sometimes moulded by shapeless plates of orthoclase, and in the same rocks reversals of order between the bisilicates and the felspars may often be noticed.

Where the felspathoids occur, their place in the order of consolidation is a variable one. These minerals usually precede the felspars, but may continue to crystallize to a later stage. The nepheline-syenites not infrequently take on a porphyritic character.

Some syenites contain basic secretions, acid veins, and other peculiarities noticed under the granites. Parallel and gneissic structures sometimes come in locally (*e.g.* Plauen'scher Grund).

Leading types. Although typical hornblende-syenites probably occur in this country (*e.g.* Malvern), very little has been written about them, and for the type-rocks we must go to foreign occurrences. The name 'syenite' as found in many of the earlier writings and maps in this country is to be understood in the old sense of hornblende-granite (including also granophyre, *etc.*), and the identification of hornblende is in very many cases erroneous. For example, the so-called 'syenites' of St David's, of Ennerdale, of Carrock Fell, *etc.*, have no claim to the title, whether the word be used in its original or its modern sense.

The rock taken as the type of *hornblende-syenite* is that of Plauen'scher Grund near Dresden (fig. 7). It is composed essentially of orthoclase, with only subordinate oligoclase, and green hornblende. Apatite, magnetite, and sphene occur as accessories, and in places a very little quartz. There is a variety in which biotite occurs in addition to the hornblende. The rock encloses dark basic secretions richer in plagioclase, hornblende, apatite, magnetite, and sphene. Further there are pegmatoid acid veins of coarse texture, in which the more basic minerals occur only sparingly, while quartz is plentiful. Almost the same description applies to other Saxon syenites, such as that of Meissen, which, however, has rather more brown mica, and further contains a little more quartz, either in grains or in micrographic intergrowth. There is rather

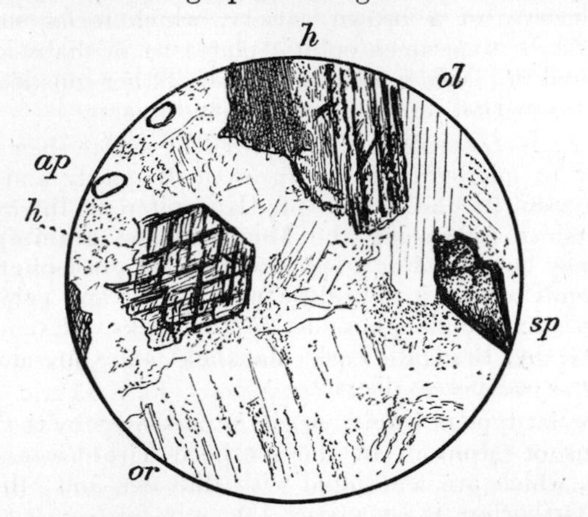

FIG. 7. HORNBLENDE-SYENITE, PLAUEN'SCHER GRUND, DRESDEN; ×20: shewing hornblende (*h*), orthoclase (*or*), subordinate oligoclase (*ol*), sphene (*sp*), and apatite (*ap*) [47].

more oligoclase than in the preceding, besides abundant sphene and apatite. A rock from Biella in Piedmont is closely similar to that of Plauen'scher Grund, sphene and apatite being rather plentiful.

Such rocks as that of Meissen may with propriety be termed *quartz-syenites*, and form a connecting link with the hornblende-granites. Again, when a triclinic felspar becomes predominant we have transitions to quartz-diorite (*e.g.* Weinheim, in the Odenwald, near Heidelberg).

The *mica-syenite* type, in which biotite predominates over hornblende, is of uncommon occurrence, except as a local variety of hornblende-syenite. More often there is some quartz present, and such rocks are found graduating into biotite-granite.

We pass on to the augite-syenites, some of which also contain more or less quartz.

The *quartz-augite-syenite* of Llanfaglen near Caernarvon has both orthoclase and plagioclase. Augite in imperfect crystals and grains is colourless in section, but tends to pass into greenish uralitic hornblende. There is original hornblende, mostly of a brown variety, which forms sometimes good crystals, sometimes ophitic plates as in diabasic rocks: it is altered in places into brown mica. Other constituents are apatite, magnetite, ilmenite, pyrites, and quartz.

Other quartz-syenites characterized by augite shew a strong tendency to micrographic intergrowth of quartz and felspar. This is seen in the larger pre-Carboniferous intrusions of Leicestershire (excepting the Mount Sorrel granite), which indeed may be classed as a less acid type of granophyre. The augite tends to pass into uralitic hornblende, and epidote is a characteristic secondary product in the rocks. Examples are seen at Groby, Bradgate Park, Markfield, and Garendon, all in the Charnwood Forest district.

A special type of *augite-syenite* is presented by the Triassic intrusions of Monzoni in the southern Tirol (monzonite of authors), which are associated with diabases and other basic rocks. Orthoclase is sometimes the only felspar, but usually there is a plagioclase in addition, forming idiomorphic crystals enclosed with the other minerals by plates of orthoclase. The augite often passes over into green hornblende, but the latter mineral also occurs as an original constituent. Biotite is usually present, in flakes sometimes earlier, sometimes later, than the plagioclase. Sphene is frequent, and zircon is often enclosed by the mica. Other constituents are apatite, magnetite, and pyrites, and in some examples a little interstitial quartz.

A peculiar augite-syenite (Laurvig type), allied in some respects to the nepheline-syenites, occurs among the Devonian intrusions of the Christiania district. While augite is usually the dominant ferro-magnesian element, it is often accompanied by biotite, ægirine, hornblende, or arfvedsonite, and the rock thus passes into mica-syenite, *etc.* Alkali-felspars (orthoclase, microcline, albite, cryptoperthite, *etc.*) make up the bulk of the rock, and are often intergrown with one another. Not infrequently they have a schiller-structure. A little quartz is rarely present; on the other hand elæolite and sometimes olivine may occur as minor accessories. The augite is occasionally green, but commonly light brown with a violet tone and slight pleochroism. Schiller-structure is common. The hornblende is green or occasionally brown, the biotite a very deep brown. The latter mineral is roughly idiomorphic, except when it is massed around magnetite or forms a marginal intergrowth with augite. The iron-ores are magnetite and sometimes hæmatite: apatite is universal, but sphene is typically absent. Zircon is a constant accessory, and sometimes builds large crystals, giving the variety 'zircon-syenite' of von Buch and other early writers. These augite-syenites are common as boulders on our East coast (Fig. 8).

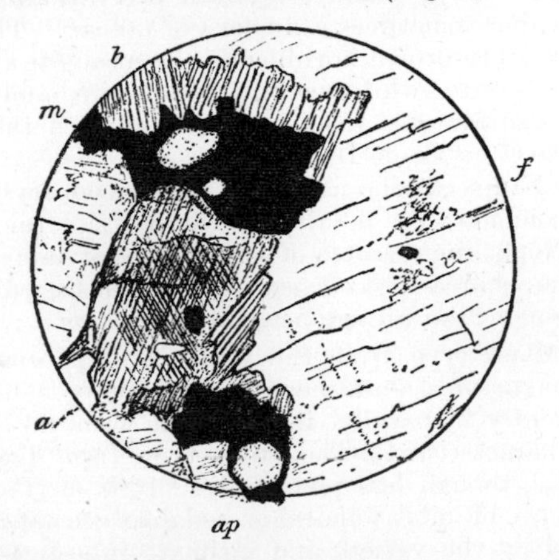

FIG. 8. AUGITE-SYENITE (LAURVIG TYPE) FROM A BOULDER ON THE YORKSHIRE COAST; ×20.

The minerals seen are cryptoperthite felspar (*f*) in large plates, augite (*a*) with schiller-structure in the interior of the crystal, deep brown biotite (*b*), magnetite (*m*), and apatite (*ap*) [1841].

The *nepheline-syenites* are in part closely allied to certain augite-syenites, and in the Christiania area, for instance, the two rocks are closely associated. The nepheline-syenite of that district (Laurdal type) differs from the Laurvig rock chiefly in the presence of elæolite and sometimes sodalite, the latter sometimes shewing a pale blue or violet tint. The same alkali-felspars as before are present, with frequent intergrowths. The abundant ferro-magnesian minerals embrace deep brown biotite, green hornblende, and light brown or purplish-brown pleochroic augite, either singly or in association. Apatite and magnetite are present, and occasionally a little olivine. A similar rock occurs in Arkansas[1], where there is also one with a porphyritic tendency and with hornblende as the chief ferro-magnesian silicate (Pulaski type).

A well-known type of nepheline-syenite is found in the Sierra de Monchique[2] in southern Portugal, where it cuts Carboniferous rocks (Foya type[3]). The proportions of elæolite and orthoclase vary; sodalite is often present, and, according to Sheibner, has sometimes a nucleus of nosean. The coloured minerals are green-brown hornblende, green augite and ægirine, and brown biotite, while apatite, magnetite, and abundant sphene are also found. Rocks comparable with this occur in Brazil, in the Cape Verde Islands, near Montreal, *etc.*[4] In some of the Brazilian rocks the nepheline is in good crystals: sodalite is not common: the dominant coloured mineral is a green ægirine. Nepheline-syenites of Cretaceous or later age occur in the Pyrenees area (Pouzac, *etc.*), and are of the same general type as those of Silurian age near Montreal, *etc.*

In the Miask type, from the Urals, a deep brown mica is the ferro-magnesian constituent. Plagioclase is often abundant, frequently in parallel intergrowth with the orthoclase. Zircon is characteristic. The Ditro type, from Transylvania, also has mica, though less plentifully. It is distinguished by its abundance of allotriomorphic sodalite accompanying the elæolite and by the variety and intimate intergrowths of the felspars, which include microcline as well as orthoclase and oligoclase. Cancrinite, sphene, zircon, and perofskite also occur[5].

In some nepheline-syenites soda- predominates over potash-felspar. A type from Litchfield[6] in Maine illustrates this.

[1] On this and other remarkable rock-types, see J. F. Williams, *Igneous Rocks of Arkansas*, vol. ii of *Ann. Rep. of Geol. Surv. Ark.* for 1890.

[2] Sheibner, *Q. J. G. S.* (1879) xxxv, 42–47.

[3] This is Blum's 'foyaite,' a name which has sometimes been used as synonymous with nepheline-syenite.

[4] See *M. M.* (1889) viii, 168; (1892) x. 42; *G. M.* 1891, 216 (*Abstracts*).

[5] For coloured figures of these rocks see Fouqué and Lévy, *Min. Micr.* Pl. XLV, fig. 1.

[6] Bayley, *Bull. Geol. Soc. Amer.* (1892) iii, 235–241.

Albite constitutes about half of the rock, the other minerals being orthoclase, microcline, elæolite, sodalite, cancrinite, a deep green biotite (lepidomelane), and a little zircon. Plagioclastic nepheline-syenites are also described from Arkansas[1]; these are ægirine-bearing rocks.

A *sodalite-syenite*, with little or no elæolite, seems to be an uncommon type. It has been found in Montana[2]. Another peculiar rock (Taimyr type), described by Chrustchoff[3] from northern Siberia may be styled *nosean-syenite*, consisting essentially of nosean and anorthoclase with some brown hornblende, biotite, zircon, *etc.* Altered *leucite-syenites*, containing pseudomorphs of orthoclase and elæolite in the form of leucite, have been described in Brazil[4], Arkansas[5], *etc.* They commonly shew porphyritic structure, and are perhaps more appropriately placed among intrusive types.

Although no genuine elæolite-bearing syenite is yet known from this country, a closely allied rock has been described by Mr Teall[6] from Loch Borolan in Sutherland. There the usual type consists essentially of orthoclase, a brown melanite garnet, and a green or green-brown biotite. A green monoclinic pyroxene is present in many examples: a brown pleochroic sphene, apatite, and magnetite occur as accessories. Nepheline is supposed to be represented by an alteration-product which often forms micrographic intergrowths with the orthoclase in patches, giving a pseudo-porphyritic aspect to the rock. Another substance, in confused aggregates with a bluish tint in reflected light, is probably one of the sodalite minerals. It is found especially in certain pegmatoid veins in the rock, consisting chiefly of orthoclase. The above rock, to which the name borolanite has been given, differs from certain garnetiferous nepheline-syenites[7] in having dominant orthoclase instead of elæolite.

1 J. F. Williams, *l.c.*, 136–140.
2 Lindgren, *Amer. Journ. Sci.* (1893) xlv, 290–297.
3 *M. M.* x, 259, 260 (*Abstr.*).
4 Derby, *Q. J. G. S.* (1891) xlvii, 254–263.
5 J. F. Williams, *l.c.* 267–277.
6 *Trans. Roy. Soc. Edin.* (1892) xxxvii, 163–178, Plate.
7 Cf. J. F. Williams, *l.c.*, pp. 229–231.

CHAPTER IV.

DIORITES.

THE diorites are plutonic rocks of medium to coarse texture, consisting essentially of a soda-lime felspar and hornblende, with less important constituents. The family so defined cannot be regarded as a natural one, its members ranging in chemical composition from sub-acid to thoroughly basic. The gabbros (characterised by a pyroxene in place of hornblende) also include intermediate as well as basic rocks, and the distinction between the hornblende- and augite-bearing types is rather an artificial one. It was established before the strong tendency of augite to pass over into hornblende was thoroughly appreciated: later research has shewn the certainty of some, and the possibility of many, of the rocks that have been termed diorites being really amphibolized pyroxenic rocks.

The more acid diorites contain free silica (*quartz-diorites*), and, except for the smaller proportion of quartz and the nature of the felspars, do not differ much from the hornblende-granites. They may have biotite in addition to hornblende (*quartz-mica-diorites*), or in some cases augite. In the *diorites* proper, without quartz, mica is not common, but the hornblende may be accompanied by augite or sometimes enstatite. The hornblende is more abundant relatively to the felspar than in the preceding types, and some of the more basic diorites consist chiefly of hornblende. These are the 'amphibolites' of some authors[1]. In some types olivine enters as a constituent (*olivine-diorites*).

[1] For a hornblende-rock (local modification of a diorite) see Fouqué and Michel Lévy, Pl. XXIII.

The occurrence of felspathoid minerals in dioritic rocks seems to be very exceptional. The *theralites* of Rosenbusch may be regarded as nepheline-diorites and nepheline-gabbros, but very little is yet known of such rocks.

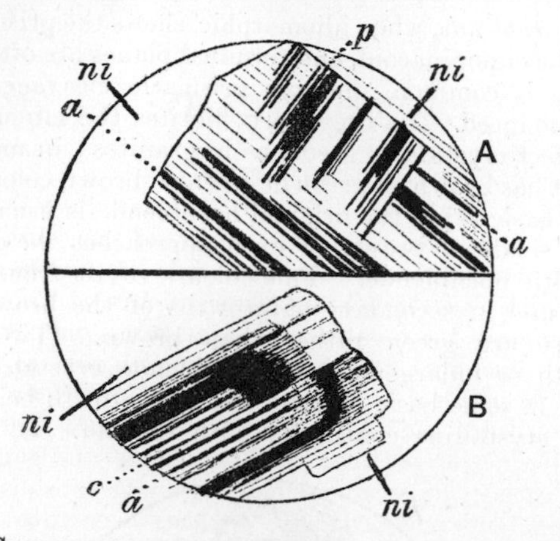

FIG. 9. CRYSTALS OF PLAGIOCLASE FELSPAR IN QUARTZ-MICA-DIORITE, BEINN NEVIS; ×20.

Crossed nicols: the vibration-planes of the nicols are indicated by the lines (*ni*) [397]. *A* shews the association of twin-lamellation on the albite (*a*) and pericline (*p*) laws. *B* shews carlsbad twinning (*c*) combined with albite-twin-lamellation (*a*) and with zonary banding.

Constituent Minerals. The felspar of the diorites is *oligoclase*, *andesine*, or *labradorite*, or exceptionally a more basic variety. The twin-lamellation on the albite type is often accompanied by pericline- or carlsbad twinning (fig. 9 *A*). In the quartz-diorites especially, the crystals frequently shew between crossed nicols a marked zonary banding, the central and marginal portions of a crystal often giving widely different extinction-angles, and the successive layers growing more acid from within outwards (fig. 9 *B*). In natural light the zones of growth may be indicated by the disposition of fluid-pores, minute scales of hæmatite, or other inclusions. The crystals are often clouded by a fine dust (probably kaolin), and may also furnish by their alteration scales of colourless mica (paragonite ?), grains of epidote, calcite, *etc.* A little *orthoclase* may be present as an accessory, behaving in the quartz-diorites as in granites, while in typical diorites it occurs interstitially.

The *hornblende*, when idiomorphic, shews the prism-faces and usually the clinopinacoid, and terminal planes are often present. Twinning is common, and the prismatic cleavage is always well pronounced. In the quartz-diorites the mineral, usually in imperfect crystals, is green, as in granites ; in more normal diorites it has brownish-green or greenish-brown colours; and in the most basic types the original hornblende is usually of some greenish shade of brown, or even approaches the deep brown of 'basaltic hornblende.' Pale colours result from bleaching, or are found in secondary outgrowths of the brown crystals, and these are green rather than brown. Two kinds of outgrowth or enlargement of hornblende crystals are to be observed in some basic diorites, the new growth being in both cases in crystalline continuity with the old. In one case a

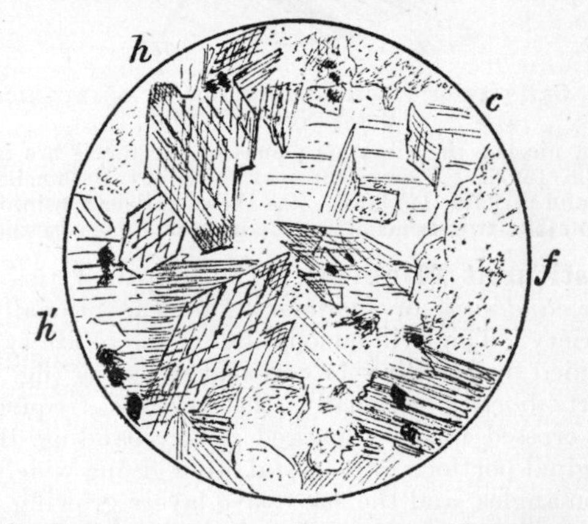

FIG. 10. BASIC DIORITE, LLYS EINION, NEAR LLANERCHYMEDD, ANGLESEY ; ×20.

The original idiomorphic brown hornblende has an extension of green hornblende on the clinopinacoid faces (h) and also a secondary fibrous outgrowth on the terminal planes (h'). The felspar (f) is much decomposed, and crystalline calcite (c) has been produced [539].

growth of green hornblende takes place on the clinopinacoid faces so as to extend the crystal, with idiomorphic contour, in the direction of the orthodiagonal: in the other case pale green or colourless hornblende grows so as to extend a crystal in the direction of its length, and may present new crystal-faces, or abut on another crystal, or frequently terminate in a ragged fibrous fringe. The second type of outgrowth at least is of secondary origin, and is formed at the expense of other minerals (fig. 10). Besides more usual types of alteration[1], the brown hornblende of diorites may shew bleaching, with separation of magnetite, or it may be converted into a brown mica or into green blades of actinolite.

The deep brown *biotite* of the diorites occurs in idiomorphic flakes, or sometimes intergrown with hornblende. It is usually not rich in inclusions. It becomes green only by partial decomposition.

The rhombic pyroxene found in a few diorites is a variety poor in iron (*enstatite*) and is usually converted into pseudomorphous pale bastite.

When *augite* is present, it is of a variety sensibly colourless in slices. If idiomorphic, it shews the octagonal cross-section due to equal development of the pinacoids and prism faces, with good prismatic cleavage and not infrequently lamellar twinning parallel to the orthopinacoid. A not uncommon feature in diorites is a parallel growth of augite and hornblende, a crystal-grain of the former mineral constituting a kernel, round which a shell of brown hornblende has grown, and this seems to occur specially in the neighbourhood of grains of iron-ore. This must be distinguished from another phenomenon frequent in the augite-bearing diorites, *viz.* the conversion of augite into brown hornblende as a secondary change. This process usually begins at the margin of a crystal or grain, but proceeds irregularly, shewing a very intricate boundary between the two minerals and often ragged scraps of one enclosed by the other. When the conversion is complete, the secondary hornblende can be distinguished from original only by inference, as, *e.g.*, when it shews the external

[1] Zirkel, *Micro. Petr. Fortieth Parallel*, Pl. III, figs. 2, 3, 4.

form of augite. In both phenomena the augite and hornblende have their plane of symmetry and longitudinal axis in common, and in longitudinal sections both extinguish on the same side of the axis.

The *quartz* of quartz-diorites has the same general characters as that of granites.

The *olivine* which occurs in some basic diorites is often in rather rounded crystals moulded by the hornblende. It is easily recognised by its high refractive index and very strong double refraction. The mineral is readily altered into serpentine, carbonates, and especially pale fibrous amphibole, the last often grown in crystalline continuity with adjacent original hornblende.

Among the iron-ores *magnetite* is the most usual, but *ilmenite* is also found. Common accessories in some varieties are *zircon* and *sphene* in characteristic crystals. *Apatite* is general, and in some basic diorites abundant: in the coarse-grained rocks it sometimes builds rather large prisms.

Structure. The structure of the dioritic rocks is variable. In the quartz-diorites the mutual relations of the minerals are those noticed in granites, though sometimes a part of the felspar has crystallized before the ferro-magnesian minerals. A micrographic intergrowth of quartz and felspar is not infrequent. Many of the quartzless diorites also follow what may be called the normal order of crystallization. Rosenbusch points out that the most marked pauses in the process of consolidation have occurred before the separation of the ferro-magnesian minerals and after that of the plagioclase; so that while the apatite, sphene, *etc.*, and the plagioclase may be markedly idiomorphic, the hornblende, biotite, and augite tend to occur in much more irregularly shaped crystals. When a miarolitic structure results from the tendency to idiomorphism in the latest crystallized elements, it is commonly obscured by the cavities becoming filled by calcite and other secondary products.

A different type of structure, though connected by transitions with the preceding, is found in many dioritic rocks. Here the plagioclase has crystallized earlier, or at least ceased to crystallize earlier, than the bisilicates; so that the dominant felspar presents idiomorphic outlines to the hornblende and (if present) augite. These latter may wrap round, or even enclose, the felspar crystals, giving an 'ophitic' structure identical with that described below as characteristic of the diabases, and the hornblendic rocks exhibiting this character have sometimes been termed hornblende-diabases. Such a structure is found more or less markedly in many of the more basic diorites, and is especially common in rocks in which the hornblende is in great part derivative after augite, but original hornblende moulding felspar is also found[1].

A porphyritic structure is not common in true diorites, but may come in as a marginal modification of a boss or stock, the porphyritic elements being crystals of hornblende or felspar (*e.g.* Lac d'Aydat in Puy de Dome).

As a special type of structure may be mentioned the orbicular (in the so-called corsite or napoleonite), where the bulk of the rock consists of spheroidal growths. These have a radial structure and consist of concentric shells composed essentially of hornblende and felspar in alternation.

Leading Types. The *quartz-mica-diorite* of the Adamello Alps, on the border of Italy and the Tirol (Tonale type) comes very near in characters to some granites[2], and has also points in common with the Monzoni syenites. The dominant felspar is a striated plagioclase, often shewing zonary banding and with a strong tendency to idiomorphic outlines; but there is frequently clear orthoclase in addition, in irregular crystal plates moulded on or enclosing the triclinic felspar. Biotite is the most constant coloured element, but hornblende is also abundant. The mutual relations of the two are variable, and both may enwrap the plagioclase. Interstitial quartz is abundant; patches of magnetite are often prominent; and zircon in little well-built prisms is general.

More typically intermediate *quartz-diorites* occur in Hun-

[1] *Q. J. G. S.* (1888) xliv, 450–453.

[2] This is the '*tonalite*' of vom Rath. Since it is an extreme type, and is classed by some petrologists with the granites, it is confusing to extend this name, as some writers have done, to all the quartz-diorites.

gary (Banat type), where they are of Tertiary age, and are the plutonic equivalents of some of the andesitic lavas. Here quartz is less abundant, and orthoclase usually absent. The characteristic zonary banding of the felspars is strongly marked. Green or brown-green hornblende is the dominant coloured mineral, but brown biotite is also common, and the two are sometimes intergrown. Crystals of magnetite and other minor accessories are found. Some varieties of the rock tend to develope a porphyritic structure.

Further examination will probably shew that some of the Scottish Carboniferous 'granites' are better classed as quartz-diorites (usually with mica). The zoned plagioclase crystals, the interstitial quartz, and other features are well exhibited (Beinn Nevis, *etc.*). There may be a rough micrographic structure. Quartz-diorites and quartz-mica-diorites occur about Garabal Hill near the head of Loch Lomond[1], and shew interesting gradations, on the one hand into granite, and on the other into quartzless diorites (mica-diorite, augite-diorite, *etc.*). Other quartz-diorites, usually with biotite as well as hornblende have been described from Arran, Glen Tilt, *etc.* The Arran rocks are believed to be of Tertiary age.

In Wicklow, east of Rathdrum, occur quartz-diorites and quartz-mica-diorites, which seem to approach granites in their characters[2]. Subordinate orthoclase accompanies the dominant triclinic felspar. The other minerals are pale green hornblende, ragged flakes of biotite, abundant quartz, apatite, and sometimes a little colourless augite (salite or malacolite). The augite-diorites, which are a common type in Wicklow, sometimes have interstitial quartz in addition to the plagioclase, hornblende, and idiomorphic salite which are their essential constituents.

Of the *simple diorites*, without quartz, good examples, probably of Carboniferous age, are found in Warwickshire and other parts of the Midlands. In the rock of Atherstone, Hartshill, the brown hornblende is in part idiomorphic towards the turbid felspar, but part of it, on the other hand, is derived from a colourless augite, and a kernel of the latter mineral

[1] Dakyns and Teall, *Q. J. G. S.* (1892) xlviii, 104–120.
[2] Hatch, *G. M.* 1889, 262, 263.

sometimes remains unchanged. Grains of magnetite are present and abundant prisms of apatite (fig. 11). Allport[1] also mentions pseudomorphs of calcite, *etc.*, after olivine. The same writer describes a fine-textured diorite from Marston Jabet, in which idiomorphic brown hornblende is set in an aggregate of triclinic felspar. Rather coarse-grained diorites are met with in the curious complex of igneous rocks in the

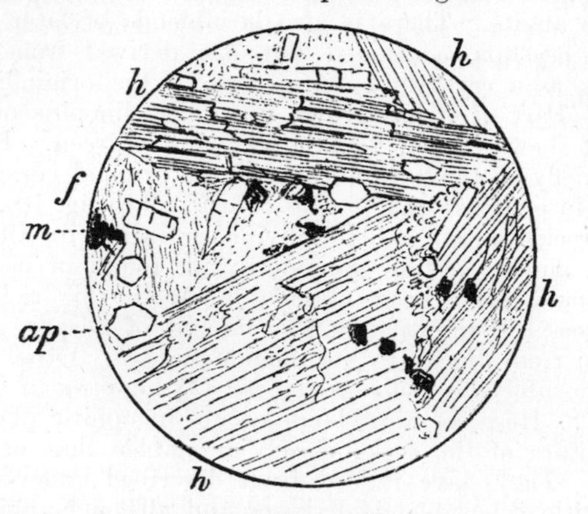

FIG. 11. DIORITE, ATHERSTONE, WARWICKSHIRE ; ×20.

The figure shews idiomorphic hornblende (*h*), turbid felspar (*f*), magnetite (*m*), and rather abundant prisms of apatite (*ap*). Cross-sections of the last shew the hexagonal shape, and longitudinal sections shew the cross-fracture [1608].

Malvern district. A specimen taken near the New Reservoir consists essentially of idiomorphic greenish-brown hornblende and labradorite felspar. The latter shews albite- and pericline-lamellation, and its decomposition has given rise to zeolites and paragonite mica. In the well known diorite of Brazil Wood[2] in Charnwood Forest, Leicestershire, the hornblende tends to be moulded on the felspar, and this departure from the granitic type of structure is observable in some other diorites from the Midland counties.

[1] *Q. J. G. S.* (1879) xxxv, 637–641.
[2] Hill and Bonney, *Q. J. G. S.* (1878) xxxiv, 224.

Various diorites occur in the interior of Anglesey. One between Gwindu and Llanfaelog is a coarse-textured rock consisting of greenish brown hornblende and turbid felspar with magnetite and apatite. The minor intrusions near Llanerchymedd[1] are of a rather different type. Brown hornblende occurs in well formed crystals and also in shapeless plates which can sometimes be seen forming at the expense of a colourless augite. There is also hornblende of later growth than the crystals mentioned but not derived from augite. It occurs as a crystalline outgrowth of the original brown crystals. Part of it has grown upon the clinopinacoid faces and itself shews crystal boundaries; this is green. Part has grown chiefly on the terminations of the original crystals and filled up interstices: this is pale or colourless (fig. 10). Some of these rocks contain a little olivine, or rather its alteration-products, and in certain specimens, not found in place, this mineral must have been abundant. With this richness in olivine goes a diminution in the amount of felspar, giving a transition from diorite to hornblende-picrite[2]. Other olivine-bearing hornblendic rocks occur near Clynog-fawr in Caernarvonshire[3]. Here the hornblende occurs in ophitic plates, and the structure of the rocks closely resembles that of typical diabases. They have indeed been described under the provisional title of hornblende-diabases, and, although augite is not often seen in them, it is possible that much of the hornblende is derivative after that mineral. The same remark applies to certain rocks at Penarfynydd[4] in the Lleyn peninsula, where both ophitic and idiomorphic augite may be seen partly converted into brown hornblende. Olivine seems to have been rare in these rocks, but they are closely associated with a hornblende-picrite rich in that mineral. Some thoroughly basic dioritic rocks, very like those of Anglesey, occur in the Lake District, *e.g.* at Little Knott[5], White Hause, and Great

[1] *G.M.* 1887, 546–552. Other types of dioritic rocks from Central Anglesey are described by Mr Blake, *Rep. Brit. Assoc.* for 1888, 403–406.

[2] Bonney, *Q.J.G.S.* (1881) xxxvii, 137–139; (1883) xxxix, 254–256.

[3] '*Bala Volc. Ser. Caern.*' 102–106.

[4] *Ibid.* 92–97.

[5] Bonney, *Q.J.G.S.* (1885) xli, 511–513, Pl. XVI, fig. 2.

Cockup[1] in the Skiddaw district. The rock at the first-named locality shews beautifully the pale fringes of hornblende which form a crystalline outgrowth of the original idiomorphic crystals. These fringes are clearly secondary, and occupy the place of destroyed felspar, *etc.* Some olivine has been present in some specimens. These Welsh and Cumbrian dioritic rocks occur usually in small laccolitic intrusions, probably of Ordovician age.

In the Isle of Man several small masses of diorite are found on Langness. The hornblende, of a greenish brown tint, is perfectly idiomorphic, but often shews secondary outgrowths. The felspars are much decomposed. Abundant zoisite, epidote, calcite, *etc.*, have been produced, and the quartz which is always found is probably all secondary. Apatite is plentiful, but a little pyrites is usually the only iron-ore present.

A *mica-diorite*, without quartz, is not a common type. It is found as a local modification of biotite-granite between Carrick Mt. and Arklow, in Wicklow. Mr Teall[2] describes a good example from Pen Voose in the Lizard district, Cornwall. This consists essentially of felspar and a reddish brown mica with only quite subordinate green hornblende and accessory sphene.

The diorites of the Scottish Highlands are not yet described in any detail. Those of the Garabal Hill district include mica-diorite and augite-diorite. The pale green augite is usually in allotriomorphic grains irregularly bordered by green hornblende. Diorites, with other hornblendic rocks, occur near Inchnadamff in Sutherland[3]. Here the hornblende is in unusually perfect crystals.

A number of dioritic rocks may be studied in the Channel Islands. A very fresh rock from the quarries of Delancy Hill, Guernsey, is an *augite-diorite*, with colourless augite as well as brown original hornblende. The latter mineral is moulded on the felspar prisms, and often borders the augite

[1] Postlethwaite, *Q. J. G. S.* (1892) xlviii, 510.
[2] Teall, Pl. XXXII, fig. 1; XLVII, fig. 3.
[3] Teall, *G. M.* 1886, 346-353.

with the usual crystallographic relation (fig. 12). A specimen from Rope-walk Quarry is also an augite-diorite with diabasic characters. The colourless augite is partly in rounded grains enclosed by the felspar, partly in shapeless plates, and the brown hornblende, apparently an original mineral, is clearly of posterior consolidation to the felspar. Magnetite is

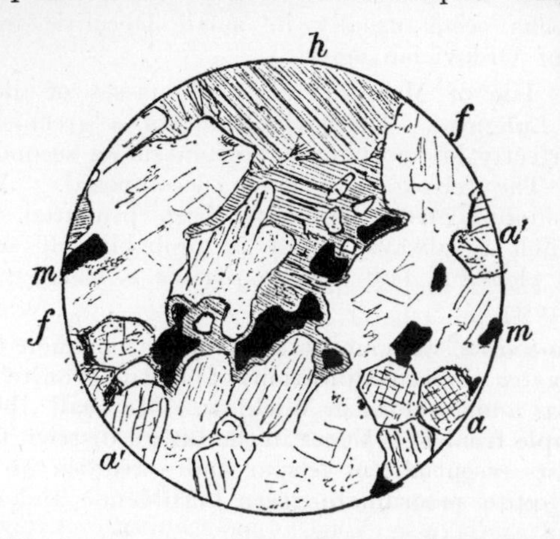

FIG. 12. AUGITE-DIORITE, DELANCY HILL, GUERNSEY; ×20.

The augite shews either sharp octagonal cross-sections (*a*) or more rounded contours (*a'*). Hornblende (*h*), magnetite (*m*), and clear plagioclase felspar (*f*) are the other constituents. Much of the hornblende occurs in marginal intergrowth with the augite, interposed between the latter mineral and the magnetite [431].

plentiful, and there are some large crystals of a rhombic pyroxene replaced by bastite. An augite-diorite from Fort Touraille in Alderney gives evidence of the conversion of augite into hornblende. Some deep brown biotite is also present, and a little interstitial quartz is the last product of consolidation.

CHAPTER V.

GABBROS AND NORITES.

THE gabbros and their allies are holocrystalline rocks, typically of plutonic habit, in which the essential constituents are a lime-soda-felspar and a pyroxene. Of intermediate to thoroughly basic character, they correspond in some sense to the diorites; but the more acid, and especially the quartz-bearing types are less represented in the pyroxenic than in the hornblendic series. According to the dominant pyroxene, we recognise *gabbro* proper (euphotide of Haüy) with diallage or augite, and *norite* (also called hypersthenite[1] or hyperite) with a rhombic pyroxene. A few of the more acid rocks contain free silica (*quartz-gabbro* and *quartz-norite*). In most of the more basic varieties olivine becomes a characteristic mineral (*olivine-gabbro* and *olivine-norite*). The majority of the rocks in this family contain more or less olivine, and the mineral may be present or absent in different specimens of the same mass.

The gabbros and norites, indeed, shew considerable variations in mineralogical constitution in parts of one mass, and most of the special types are probably to be regarded as only local modifications. Thus, by the failure of one or other of the chief constituents of a gabbro, we may have an almost pure *felspar-rock* (labrador-rock, anorthosite) or *pyroxene-rock*

[1] In many of the 'hypersthenites' of the older writers the supposed hypersthene is only a highly schillerized diallage.

(diallage-rock, *etc.*, pyroxenite[1] of Williams). By the disappearance of the pyroxene of an olivine-gabbro, we have the so-called *troctolite* (Ger. Forellenstein), composed essentially of felspar and olivine: with abundant olivine and diminishing felspar we have transitions to the succeeding family of peridotites.

The name *hornblende-gabbro* has been used for rocks of this family which contain hornblende in addition to pyroxene, or in which original pyroxene is more or less completely replaced by hornblende[2]. When the conversion is complete we have no decisive criterion for verifying the derivative nature of the hornblende, and, as already remarked, the distinction between diorite and gabbro is a somewhat artificial one[3].

A historical account of the classification of the gabbros and allied rocks has been given by Bayley[4].

Constituent minerals. The felspar of the gabbros and norites ranges in different examples usually from *labradorite* to *anorthite*. It builds large irregularly-shaped plates with, as a rule, rather broad[5] lamellæ (albite twinning), often crossed by fine pericline-striation. Zonary structure is typically not found. Besides fluid-pores and inclusions of earlier products of crystallization, the felspars often shew more or less marked schiller-structure (fig. 1, *g*). The modes of alteration of the felspars are various: Rosenbusch notes the curious fact that calcite is seldom formed. The 'saussurite' change seems to belong to dynamic metamorphism rather than weathering (see below, Ch. XXI). Any plagioclase more acid than labradorite is exceptional, and so is the occurrence of *orthoclase* (*e.g.* Lake Superior region[6]).

[1] *Amer. Geol.* (1890) vi, 40–49. Williams regarded the pyroxenites as a group coordinate with the peridotites. The name is ill-chosen, having been employed in two or three other quite different senses.

[2] R. D. Irving, *Copper-bearing Rocks of L. Superior*, 56–58, Pl. VII.

[3] Prof. Cole restricts the name gabbro to the olivine-bearing (corresponding roughly to the basic) division, and styles the intermediate felspar-augite-rocks 'augite-diorite.'

[4] *Journ. of Geol.* (1893) i, 435–456.

[5] Cut nearly parallel to the brachypinacoid, such a crystal appears untwinned.

[6] R. D. Irving, *Copper-bearing Rocks of L. Superior*, 50–55, Pl. V, VI.

The *augite* of the gabbros builds irregular crystal-plates and wedges of very pale green or light brown colour. Besides the usual prismatic cleavage, an orthopinacoidal cleavage and *diallage*-structure are very common. Instead of this, there is sometimes a very minute lamellar twinning parallel to the basal plane (*e.g.* Carrock Fell). The common twin parallel to the orthopinacoid is often associated with this (fig. 13 *A*). De-

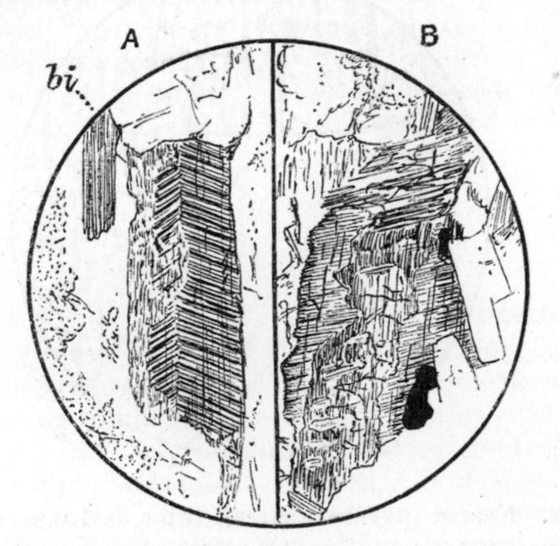

FIG. 13. PYROXENES IN THE GABBRO OF CARROCK FELL, CUMBERLAND; ×20.

The dominant variety is an augite with basal striation. *A* shews this structure combined with twinning on the orthopinacoid to give the 'herring-bone' structure. The mineral is partly converted to green hornblende [1870]. *B* shews a parallel intergrowth of the augite with enstatite, the latter mineral forming the core and the former the outer shell, but with detached portions of augite enclosed in the enstatite in micrographic fashion [2279].

composition gives a scaly or fibrous aggregate of chlorite and serpentine with other products. Another common alteration is the conversion to hornblende, which may be light green and fibrous (uralite) or deep brown and compact.

The rhombic pyroxenes, *bronzite* and *hypersthene* occur in rather rounded but idiomorphic crystals. A schiller-structure[1]

[1] Rosenbusch-Iddings, Pl. VII, fig. 5.

is common in many norites and gabbros (fig. 14). The most usual alteration is into distinct pseudomorphs of the serpentinous mineral bastite. This is pale green or yellowish with slight pleochroism and low polarization-tints. The pseudomorph is built of little fibres arranged longitudinally, and

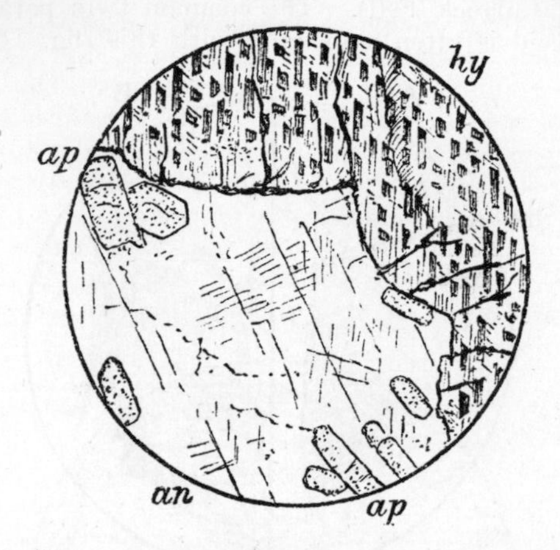

FIG. 14. NORITE (HYPERSTHENITE), COAST OF LABRADOR; ×20.

Consisting of hypersthene (*hy*), felspar (*an*), and apatite (*ap*). Schiller-inclusions are strongly developed in the hypersthene and to a less extent in the felspar [G 444].

is traversed by irregular cracks which the fibres do not cross (see fig. 20). The individual fibres give straight extinction, but, as there is a slight departure from perfect parallelism in their arrangement, a very characteristic appearance is offered. The rhombic pyroxenes also shew uralitization.

In the rocks here included original *hornblende* occurs only as an occasional accessory: a deep brown variety occurs in some norites. Brown *biotite* may also occur as a minor accessory (*e.g.* Carrock Fell; St. David's Head), and it may be intergrown with augite (Stanner Rock)[1].

When *olivine* is present, it builds imperfect crystals or

[1] Cole, *G. M.* 1886, p. 221, fig. 3.

rounded grains, colourless in slices. Where it adjoins felspars, it is often bordered by a rim of hypersthene. The olivine sometimes has schiller-inclusions.

The characteristic mode of alteration of olivine is 'serpentinization.' This process begins round the margin of the crystal-grain and along the, usually irregular, network of cracks which traverses it. Along these, as a first stage, strings of granular magnetite separate out. The immediate walls of the cracks are converted into pale greenish or yellowish fibrous serpentine, the fibres set perpendicularly to the crack, and giving straight extinction and low polarisation-tints. At this stage the meshes of the network are occupied by unaltered remnants of olivine. These may be subsequently altered to serpentine, which is of a different character from that first formed, being often sensibly isotropic[1]. As a last stage, some

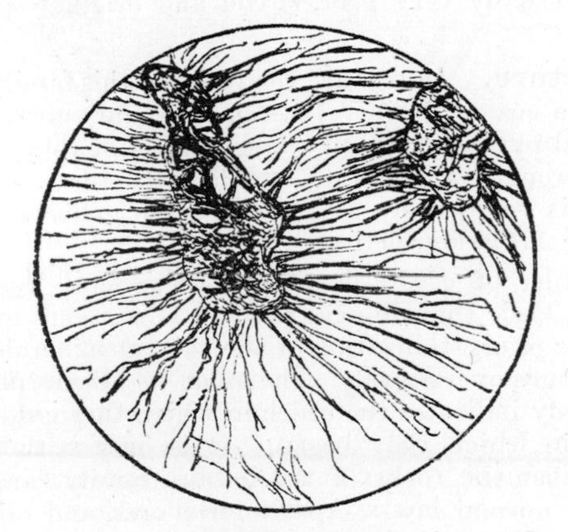

FIG. 15. LABRADORITE-OLIVINE-ROCK (TROCTOLITE), COVERACK COVE, CORNWALL; ×20.

The olivine is almost wholly converted into serpentine (a few clear granules remaining), and the consequent expansion has caused radiating fissures through the surrounding felspar [1116].

[1] This effect is possibly due to the overlapping of a crowd of minute fibres or scales without any definite orientation. For successive stages of serpentinization of olivine, see Geikie, p. 174, fig. 33.

of the magnetite may be reabsorbed, giving a deeper colour to the serpentine pseudomorph. The change from olivine to serpentine involves an increase of volume, which gives rise to numerous radiating cracks traversing adjacent minerals. These cracks are injected with serpentine, usually isotropic (fig. 15).

Where original *quartz* occurs in gabbros, *etc.*, it has the same properties as that in granites. Usually it forms part of a micrographic intergrowth.

Original iron-ores occur only sparingly in some rocks of the gabbro family, but sometimes become abundant. They are *ilmenite* (with leucoxene as a decomposition-product) and *magnetite*. In some cases brown grains of *picotite* are found. The *apatite* builds the usual hexagonal prisms or sometimes short rounded grains (fig. 14). In other accessories the rocks are usually very poor, zircon and original sphene being absent.

Structure. In texture the rocks of this family vary from medium to coarse grain. In some the individual crystals of felspar and pyroxene attain a large size, and they are then, as a rule, strongly affected by schiller-structures. Porphyritic structure is very rarely met with in the gabbros and norites (Skye and Ardnamurchan).

The order of crystallization is in general less decisively marked in basic than in acid rocks. This seems to be due to the periods of crystallization of the several minerals having in great measure overlapped. The relative idiomorphism of the crystals only indicates the order in which they *ceased* to form, not that in which they began. It is only with this understanding that the rocks of the gabbro family can be said to follow the normal law. Apatite, iron-ores, and olivine, when present, are the earliest minerals and are clearly idiomorphic, while in the special types containing orthoclase and quartz these minerals have always crystallized last. But as regards the two main constituents, augite and plagioclase, the mutual relations are not always the same. In many gabbros the felspar is more or less distinctly embraced by the augite or diallage, but if this character becomes marked there are always other features which indicate a transition to the diabase type.

The more typical gabbros are often thoroughly hypidiomorphic ; or the augitic constituent, especially if very abundant, may be embraced by the felspar. When a rhombic pyroxene enters, it is idiomorphic towards the monoclinic, and usually towards the felspar also.

In many plutonic rocks there is an evident tendency for the earlier formed minerals to serve as nuclei round which the later ones have crystallized. This tendency is most marked in basic and ultrabasic rocks. Thus in gabbros and norites the pyroxenes often form a more or less continuous ring or

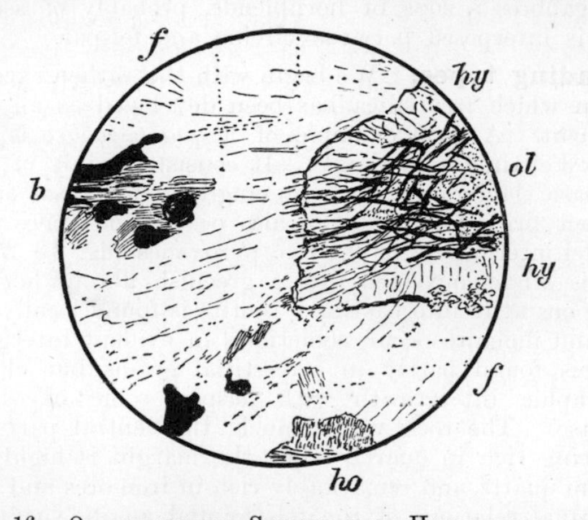

FIG. 16. OLIVINE-NORITE, SEILAND NEAR HAMMERFEST ; ×15.

A much-fissured crystal of olivine (*ol*) is surrounded by a continuous ring of hypersthene (*hy*) interposed between it and the anorthite felspar (*f*). There is a little brown hornblende (*ho*) and some brown biotite (*b*) clinging about the iron-ore grains [418].

'*corona*' round olivine or iron-ores (fig. 16). Bayley[1], while noting this feature, further describes fibrous intergrowths of felspar and augite surrounding olivine or magnetite. These seem to be original, but in other cases there is reason to believe that a mineral bordering another one is of secondary

[1] *Amer. Journ. Sci.* (1892) xliii, 515–518 ; *Journ. of Geol.* (1893) i, 702–710.

origin. Good examples are figured and described by G. H. Williams[1] in the hypersthene-gabbros of the Baltimore district. Here both hypersthene and diallage are surrounded by a double '*reaction-rim*' of hornblende, interposed between the pyroxene and the felspar and due to a reaction between them. The inner zone of the rim is of fibrous, the outer of compact hornblende. They are apparently the beginning of a process by which the pyroxenes are eventually wholly transformed into green hornblende, and the author named considers that they do not necessarily imply dynamic metamorphism. In many olivine-gabbros a zone of hornblende, probably of secondary origin, is interposed between olivine and felspar.

Leading types. We begin with the rather exceptional rocks in which free silica has been developed as an original constituent. A good example of a *quartz-gabbro* is that of Carrock Fell in Cumberland[2]. It consists mainly of a somewhat basic labradorite and an augite with basal striation. Imperfect prisms of enstatite also occur, and there is often a parallel intergrowth of the two pyroxenes (fig. 13 *B*). The augite is often converted into a greenish fibrous hornblende and the enstatite into bastite. Biotite is found locally. Magnetite and ilmenite occur, sometimes in evident intergrowths. Quartz is found partly in interstitial grains but chiefly in micrographic intergrowth with felspar, some of which is orthoclase. The rock varies much, the central part of the mass being rich in quartz, while the margin is highly basic, free from quartz, and remarkably rich in iron-ores and apatite. The mutual relations of the felspar and augite vary, but on the whole the augite tends to envelope the felspar.

Specimens of the gabbro of St David's Head, also intrusive in Lower Palæozoic strata, are identical with the rock just described, except that the highly basic modification is not found. Biotite is rather more plentiful, and the quartz and micropegmatite occur rather more sparingly. The rhombic

[1] *Bull. No.* 28 *U. S. Geol. Serv.* (1886) and Plates. See also Kemp on gabbros of L. Champlain, *Bull. Geol. Soc. Amer.* (1894) v, 217–221, with numerous references to other described cases.

[2] *Q. J. G. S.* (1894) l, 316–318, Pl. XVII. The rock has been termed hypersthenite, but the rhombic pyroxene is always subordinate to the monoclinic and sometimes wanting.

pyroxene is represented by pseudomorphs of pleochroic green bastite, always abundant. The mutual relations of the augite and labradorite vary, even in one slide: very frequently the former mineral is moulded on, or embraces, the latter. This tendency to the 'ophitic' structure, together with the absence of diallagic structure in the augite, the rather abundant occurrence of iron-ores, and other features, indicates an approach to the diabase type, which is also developed in the district. It is noteworthy that the diabase of the Whin Sill, to be noticed below, in its coarse-grained central part, takes on characters almost indistinguishable from those of the Carrock Fell and St David's Head rocks.

The well-known rocks of the Lizard district[1] in Cornwall are, for the most part, simple *gabbros* without olivine, although that mineral occurs in some varieties. Judging from the cases in which precise determinations have been made, the felspar seems to be labradorite in the less basic rocks, anorthite in the most basic. It shews broad albite-lamellæ, often crossed by others following the pericline law. The pyroxene varies from a pale green diopside, almost colourless in slices, to typical diallage, the diallagic structure being often seen to affect only part of a crystal. The enstatite-group is wanting or rare. When olivine occurs, it builds colourless grains shewing various stages of serpentinization.

The Lizard gabbros exhibit, however, numerous modifications which are ascribed to dynamic metamorphism, especially the conversion of the felspars to 'saussurite' and of the augite to amphibole. The minutely granular mineral-aggregate known as saussurite is opaque in any but the thinnest slices, and can be studied only under high magnifying powers. The change may be seen to begin in spots in the felspar crystals and spread to the whole. The pyroxene passes over into uralitic or actinolitic or compact hornblende in different cases[2], the secondary amphibole being pale green or brown or colourless, or sometimes having a bright emerald-green colour (smaragdite). According as one or both of these changes have

[1] Teall, *G. M.* 1886, 483–485. For descriptions of particular varieties, see Bonney, *Q. J. G. S.* (1877) xxxiii, 884–915, and other papers.

[2] Teall, Pl. XVIII, fig. 2.

affected the original felspar-pyroxene-rock, we have saussurite-diallage-gabbro, felspar-hornblende-gabbro, or saussurite-hornblende-gabbro.

Another mineral considered to be of secondary origin is the rhombic amphibole anthophyllite[1]. This sometimes occurs in colourless and rather fibrous crystals forming a zone round grains of altered olivine, and surrounded in turn by an outer zone of green actinolite.

Among rocks which have been styled *hornblende-gabbro*, some examples from Guernsey (Bellegreve) exhibit very beautifully the conversion of colourless augite into brown or greenish brown compact hornblende, the process being seen in every stage. In some slides no augite remains, and, without comparison with other specimens, the rock might be taken for a true diorite, but the hornblende is probably all derivative. The ferro-magnesian silicates are often moulded on the felspar, which is of a basic variety. Magnetite and apatite are the only other constituents. The transformation of augite to hornblende is seen in many other gabbros, *e.g.* those of Cornwall mentioned above.

The Tertiary gabbros of the western islands of Scotland, described by Prof. Judd[2], are in general *olivine-gabbros*. They consist essentially of felspar, augite, and olivine. The felspar seems to be typically labradorite, but varies in different examples, even to anorthite. The augite may or may not shew a diallagic character. This and the olivine are sometimes of varieties rich in iron, and give rise by alteration to magnetite, which does not usually occur as an original mineral. The structure is more or less typically hypidiomorphic, but if crystal-faces are developed, it is in the felspar, not the augite. A rhombic pyroxene may partly take the place of the monoclinic, giving transitions from gabbro to norite. Prof. Judd has remarked that in the deeper-seated parts of the gabbro-masses all the minerals may be affected by schiller-structures.

[1] Teall, *M. M.* (1888) viii, 119.

[2] *Q. J. G. S.* (1886) xlii, 49–89, Pl. IV. Teall, Pl. XVI, fig. 2; XVIII, fig. 1; XXV. For account of some remarkable varieties occurring in Skye, see also Geikie and Teall, *Q. J. G. S.* (1894) l, 650–655, Pl. XXVIII. The rocks there described are chiefly free from olivine, and contain plenty of original magnetite, partly titaniferous.

With 'schillerization' parallel to the orthopinacoid only, the augite becomes diallage (Mull and Rum); with two or three directions of schiller-structure, it becomes what Prof. Judd has called 'pseudohypersthene' (Skye and Ardnamurchan).

Very similar to the Scottish Tertiary gabbros are those of the Carlingford district in Ireland, probably of like age. Prof. von Lasaulx[1] described specimens consisting of anorthite, diallage, and olivine, and likened them to the gabbro of Store Bekkafjord in Norway. These were from Slieve Foy. From the neighbouring hill of Barnavarve Prof. Sollas[2] describes a gabbro free from olivine, consisting of a basic felspar (anorthite or bytownite) with rhombic and monoclinic pyroxenes which shew rather remarkable intergrowths. Here is also a variety of the rock containing interstitial micro-pegmatite, which the author named believes to be due to a later injection[3].

Among other British examples may be mentioned that of Stanner Rock near New Radnor[4], which contains biotite, partly in parallel intergrowth with the augite. This feature is found also in olivine-gabbros at Mte Monzoni in the Tirol and at Radauthal near Harzburg in the Harz, a district rich in olivine-gabbros, which pass into norites, troctolites, and peridotites. The 'black gabbro' of Volpersdorf in Silesia is an olivine-bearing type, the 'green gabbro' being free from that mineral.

As already intimated, many of the rocks in this family contain both augite (or diallage) and hypersthene in varying proportions, and no hard line is to be drawn between gabbros and norites. In Sweden the rocks termed 'hyperite' by Törnebohm vary between olivine-gabbro and norite, olivine and hypersthene appearing to replace one another, so that the total of the two remains about the same in the different varieties. The same thing is seen in the north of Norway and elsewhere. A well-known example of *norite* comes from the island Hitterö off the west coast of Norway. The rhombic pyroxene is a hypersthene rich in iron, but, as is often the

[1] *Sci. Proc. Roy. Dubl. Soc.* (1878) ii, 31–33.
[2] *Trans. Roy. Ir. Acad.* (1894) xxx, 482–486.
[3] *L.c.* 487, *etc.*
[4] Cole, *G.M.* 1886, 223–225, fig. 3.

case, the ferriferous ingredient is concentrated in numerous deep brown schiller-inclusions, leaving the general mass of the crystal pale and scarcely pleochroic. Some specimens have a considerable amount of iron-ore (probably titaniferous) surrounded by green hornblende.

The same strongly schillerized hypersthene is well exhibited by the norites of the Labrador coast (fig. 14). Patches of brown hornblende and biotite are sometimes intergrown with it. In places the hypersthene becomes bleached, with a separation of granular magnetite. The other main constituent is felspar (usually typical labradorite but sometimes a more basic variety), moulded on the imperfect crystals of hypersthene. Stout prisms of apatite also occur, and sometimes patches of iron-ore bordered by brown mica. In Britain typical norites occur in Aberdeenshire (near Ellen) and Banff.

Several varieties of hypersthene-bearing rocks have been described by G. H. Williams[1] from the Cortlandt district on the Hudson river. The norite proper consists mainly of andesine and hypersthene, both shewing schiller-inclusions. There is accessory biotite, and a curious feature is the occurrence of large crystals of orthoclase enclosing the other minerals in 'pœcilitic'[2] fashion. In other rock-types from this district the hypersthene is associated with green or brown hornblende (hornblende-norite), with biotite and magnetite (mica-norite), or green augite and biotite (augite-norite). Another rock, intermediate between norite and gabbro, is the hypersthene-gabbro described by the same author from Baltimore[3]. This rock consists of bytownite, diallage, and hypersthene, with some magnetite and apatite. It shews the 'reaction-rims' already referred to, and passes over into a 'gabbro-diorite' or hornblende-gabbro, in which the hornblende is derivative from the pyroxenic minerals.

The *felspar-rocks* known in America as anorthosite must be regarded as peculiar members of the gabbro family. Such rocks, of pre-Cambrian age, occupy extensive tracts in Minne-

[1] *Amer. Journ. Sci.* (1887) xxxiii, 135–144, 191–194.
[2] See below, p. 75.
[3] *Bull. No.* 28 *U. S. Geol. Surv.* (1886) with Plates.

sota[1], *etc.*, near Lake Superior. The felspar which makes up almost the whole of these coarse-textured aggregates varies from labradorite to anorthite in different localities. A little augite, of faint violet-brown tint in sections, is the only other original mineral, and this occurs both in grains and as minute parallel interpositions in the felspar. Similar rocks have been described by Adams[2] in the so-called Norian of several districts in Canada. In this country gabbros are known to pass only locally into labradorite-rock, *etc.*, by the failure of the pyroxenic constituent (*e.g.* Lenkeilden Cove, Lizard).

On the other hand, gabbros pass locally by the more or less complete disappearance of felspar into rocks consisting essentially of pyroxene. Diallage-rocks occur in some British districts of gabbros (Lendalfoot in Ayrshire)[3]. More striking are the pure *pyroxene-rocks* to which Williams in America has given the name 'pyroxenite.' The Webster type[4] is described from North Carolina and Maryland, and consists of a rhombic and a monoclinic pyroxene forming an even-grained crystalline aggregate. It is in fact a bronzite-diopside-rock. Another type of pyroxene-rock of similar characters comes from Fobello[5] in Lombardy. This is of somewhat coarser texture, and consists of diallage and hypersthene, the former predominating. The hypersthene shews strong pleochroism in red and green tints.

By the dwindling and disappearance of the pyroxene, olivine-gabbros pass into *felspar-olivine-rock*, known as troctolite (Ger. Forellenstein). This consists essentially of labradorite or some more basic felspar, often anorthite, with a smaller proportion of olivine, which may be more or less serpentinized (fig. 15). Such rocks are known among the 'black gabbros' of Volpersdorf in Silesia, in the Harz, and in many

[1] Irving, *Copper-bearing Rocks L. Superior*, 59–61, Pl. VII, fig. 4; Lawson, *Bull. No.* 8 *of Geol. and Nat. Hist. Surv. Minn.* 1893 (*Abstr. Min. M.* x, 263). The very coarse-textured felspar-rock of Labrador, with its beautiful schiller-structure, is in all mineralogical collections.

[2] *Rep. Brit. Assoc.* for 1886, 666, 667.

[3] Bonney, *Q. J. G. S.* (1878) xxxiv, 778–780.

[4] G. H. Williams, *Amer. Geol.* (1890) vi, 40–49, Pl. II, fig. 2. (*Abstr. M. M.* ix, 250, 251.)

[5] *G. M.* 1891, 169, 170.

other districts. In Britain we have good examples in the gabbro-district of the Lizard[1] and among the Tertiary intrusions of the Scottish islands. Prof. Judd[2] describes a fresh and rather fine-textured anorthite-olivine rock from Halival in Rum. Another example is described by Prof. Bonney[3] from Belhelvie in Aberdeenshire. This and the Cornish specimens have some small amount of diallage.

It has been noticed above that an ordinary gabbro may pass into a variety very rich in magnetite and ilmenite (*e.g.* Carrock Fell). Some gabbros and norites, in Scandinavia, in Minnesota, and elsewhere, shew very basic modifications which are almost pure *iron-ore-rocks*[4]. As a rule, they are highly titaniferous. An augite-magnetite-rock, consisting of crystal grains of augite set in a framework of titaniferous magnetite, is one of the varieties of the curious banded gabbros of Skye[5].

[1] See Teall, Pl. VIII, fig. 2.
[2] *Q. J. G. S.* (1885) xli, Pl. XIII, fig. 5.
[3] *G. M.* 1885, 441, 442.
[4] Vogt, *G. M.* 1892, 82–86 (*Abstract*).
[5] Geikie and Teall, *Q. J. G. S.* (1894) l, Pl. XXVIII. For descriptions of iron-ore-rocks from Cumberland in Rhode Is. and Taberg in Sweden see Wadsworth, *Lith. Stud.* 75–81, Pl. I, II.

CHAPTER VI.

PERIDOTITES (INCLUDING SERPENTINE-ROCKS).

THE peridotites are holocrystalline rocks of ultrabasic composition, in which felspar is typically absent and olivine is the most prominent constituent. They were separated from the more normal basic rocks by Rosenbusch, but, though their marked characters make it desirable to discuss them apart, they do not constitute a family comparable, *e.g.*, with that of the gabbros in importance. The peridotites do not usually occur in large bodies of uniform rock. In many localities they are seen to be only local modifications of olivine-gabbros, olivine-norites, or olivine-diorites, and they shew frequent transitions from one type to another.

For so small a group, a needless multiplicity of names has been created. The simple olivine-rock is the 'dunite' of Hochstetter. With the addition of enstatite we have the 'saxonite' of Wadsworth[1], 'harzburgite' of Rosenbusch; other types are styled 'lherzolite,' 'eulysite,' *etc.*, and the name 'picrite' is used for those characterised by augite or hornblende, usually with some felspar. For our purposes it will be sufficient to separate the *picrites*, rich in the bisilicate constituents and having usually subordinate plagioclase, from the more typical *peridotites*, very rich in olivine and non-felspathic. Different types may be specified by prefixes in the customary way (*e.g.* hornblende-picrite, enstatite-peridotite, *etc.*).

[1] *Lithological Studies* (1884, Camb. Mass.). This work contains many descriptions of peridotites and meteorites, with a number of useful, coloured plates.

Many of the meteorites ('stony meteorites' as distinguished from meteoric irons) have a mineral composition allied to that of the terrestrial peridotites, but often with special accessory minerals and peculiar structures.

In consequence of the unstable nature of their principal constituent mineral, the peridotites are very readily decomposed, and most of the serpentine-rocks have originated in this way.

Constituent minerals. In the typical peridotites *olivine* makes up from half to nearly the whole of the rock. If not so abundant that its crystals interfere with one another, it builds idiomorphic or rounded crystals. The mineral is colourless in thin slices, and shews either irregular cleavage-traces or a network of fissures. It often has schiller-inclusions of the nature of minute negative crystals enclosing dendritic growths of magnetite (fig. 1 *h*). Alteration along cracks gives rise to strings of magnetite granules, and complete destruction produces pseudomorphs of greenish or yellow serpentine, or sometimes colourless fibrous tremolite, *etc.*

Of the other ferro-magnesian silicates the commonest in typical peridotites is a rhombic pyroxene; either colourless or pale yellow (*enstatite*) or with faint green and rose pleochroism (*bronzite*): varieties rich in iron do not often occur. The crystals often tend to be idiomorphic. Any marked schiller-structures are not very common. Decomposition results in pseudomorphs of bastite[1]. The *augite* is either light brown to colourless, with a high extinction-angle (about 40°) as in many diabases, *etc.*, or it may shew a faint green tint (chrome-diopside). A conversion to brown hornblende is common in the picrites, and so also are parallel growths of augite and brown hornblende, the former being the kernel.

The *hornblende* may be a green or pale actinolitic variety, but in many of the picrites it is 'basaltic' hornblende with an extinction-angle of about 20° and colour varying from deep brown to colourless. The pale variety seems due to bleaching, often accompanied by a discharge of magnetite dust. The *biotite* of peridotites is also frequently of a pale tint.

[1] Fouqué and Lévy, *Min. Micr.* Pl. LIII, LIV.

Some peridotites have little octahedra of *magnetite*, but some other spinellid mineral is more characteristic. It may be *chromite* (deep brown or opaque), *picotite* (coffee-brown), or *pleonaste* (green). These minerals usually build irregular rounded grains. In some of the rocks *perofskite* is a characteristic mineral, in minute crystals.

A basic *felspar* occurs in many of the picrites, but is wholly wanting in the more typical peridotites. Some types have accessory garnet, which is always the magnesian variety *pyrope*, red-brown in slices. Metallic *nickeliferous iron* occurs in some of the meteoric peridotites, besides special minerals such as troilite.

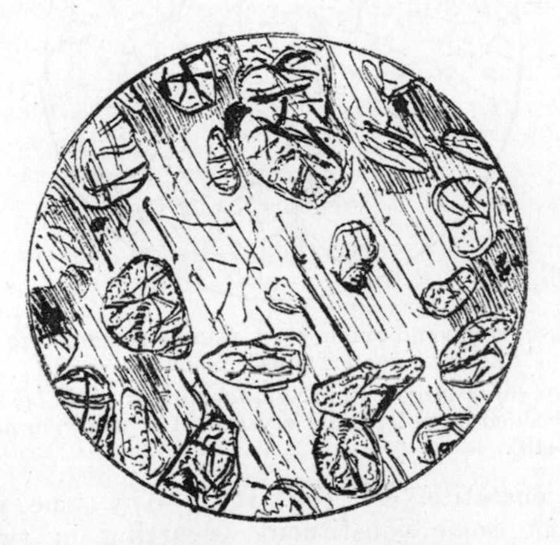

FIG. 17. PŒCILITIC STRUCTURE IN HORNBLENDE-PICRITE, MYNYDD PENARFYNNYDD, CAERNARVONSHIRE; × 20.

The large plate enclosing olivine-grains and filling the field is a single crystal of hornblende. It is mostly colourless, but becomes deep brown in capriciously arranged patches round the edge [725].

Structure. The constituents follow, as a rule, the normal order of consolidation, the olivine constantly preceding the bisilicates. In many picrites, and in other types not too rich in olivine, the more or less rounded crystals of olivine are enclosed by large plates of pyroxene or hornblende (*pœcilitic* structure[1], fig. 17). When felspar occurs, it is later than the pyroxenes, but in the hornblende-picrites it is often moulded in ophitic fashion by part of the hornblende.

In the most basic peridotites the largely predominant olivine builds a granular aggregate, in which may be imbedded, with a *pseudo-porphyritic* appearance, relatively large

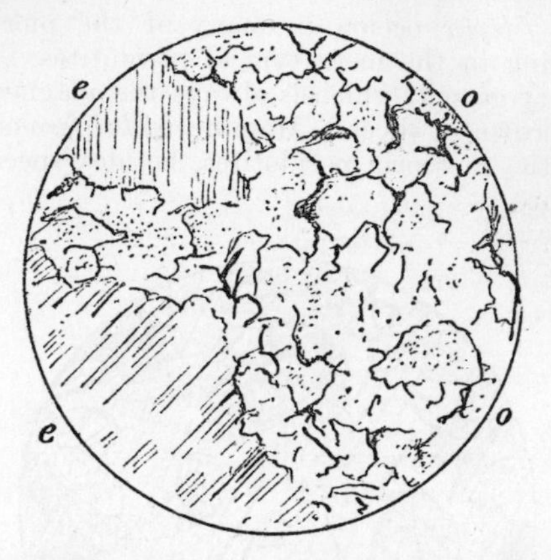

FIG. 18. ENSTATITE-PERIDOTITE WITH PSEUDO-PORPHYRITIC STRUCTURE, SKUTVIK, NEAR TROMSÖ, NORWAY; × 20.

Here olivine (*o*) is largely in excess, forming a granular aggregate in which are imbedded large irregular crystals of a yellowish partly altered enstatite (*e*) [440].

crystals of enstatite, *etc.* (fig. 18). Any true porphyritic structure (*i.e.* some constituent occurring in two distinct generations) is rare in this family of rocks, the minerals usually forming an even-grained aggregate.

The pyrope-bearing peridotites often shew a special type of structure, each garnet-crystal being surrounded by a broad border or shell known as *celyphite*[2] (Ger. Kelyphit). This border is sharply divided from the garnet, and possesses a

[1] This is quite analogous to the ophitic structure of diabases, *etc.* See G. H. Williams, *Amer. Journ. Sci.* (1886) xxxi, 30, 31; *Journ. of Geol.* (1893) i, 176.

[2] Rosenbusch-Iddings, Pl. XIV, fig. 4.

marked radial fibrous structure. The name is not applied to any particular mineral, and the so-called celyphite is not always of the same constitution. A pale or colourless augite is common, while brown hornblende and enstatite are sometimes found, and brown *picotite* frequently accompanies the pyroxene. Again, brown biotite and magnetite have been observed[1]. A celyphite-border round garnet is also a characteristic feature in pyroxene-garnet-rocks (eclogites). Some petrologists have regarded it as a secondary 'reaction-rim,' but there seems to be no decisive reason for rejecting the primary origin of the growth.

Most of the meteoric peridotites have a peculiar structure termed *chondritic*[2]. A fine-grained matrix of olivine, enstatite, chromite, *etc.*, encloses numerous round grains (*chondri*) consisting of the same minerals. In these chondri the crystals very commonly have a tendency to diverge from a point on the circumference.

Leading types. Numerous examples of rocks rich in olivine are known from the old gneiss area of Sutherland, from the western islands of Scotland, from North Wales, Cornwall, *etc.* There are frequent transitions from felspar-bearing picrites to thoroughly ultrabasic peridotites[3].

At Penarfynnydd[4], on the south-west coast of Caernarvonshire, is an Ordovician intrusion ranging from *hornblende-picrite* to a hornblende-peridotite very rich in olivine. The hornblende is either deep brown or colourless, in the same crystal, and it encloses the rounded grains of olivine with typical pœcilitic structure (fig. 17). A colourless augite and a deep brown biotite occur, with a little original magnetite. Part of the hornblende is formed at the expense of augite. Anorthite is often present, usually embraced by the hornblende. Similar rocks occur in central Anglesey, where

[1] Diller, *Amer. Journ. Sci.* (1886) xxxii, 123: *Bull. No.* 38 *U. S. Geol. Surv.* (1887) 15–17.

[2] For figures see Wadsworth's '*Lithological Studies*'; Lockyer, *Nature* (1890) xli, 306, 307.

[3] For figures of several of these rocks, see Teall.

[4] *Q. J. G. S.* (1888) xliv, 454–457. '*Bala Volc. Rocks of Caern.*' 99–101.

secondary crystal-outgrowths from the hornblende are frequent[1]. Prof. Bonney[2] has described some of these rocks, which occur as boulders on the west coast of Anglesey. The same writer has described from Sark[3] a somewhat different type in which a pale altered mica is a prominent mineral, besides pale or greenish actinolite. This seems then to be a mica-hornblende-picrite, and Prof. Bonney compares it with the Scye type mentioned below. G. H. Williams[4] has given an interesting account of hornblende-picrites from the Cortlandt district on the Hudson River. They resemble very closely the British examples and a well-known rock from Schriesheim, near Heidelberg, the bleaching of the brown hornblende and subordinate brown biotite being a characteristic feature. Williams uses the name 'cortlandtite' for these rocks, and they may conveniently be styled the Cortlandt type.

An *augite-picrite* of Carboniferous age is found at Inchcolm[5], near Edinburgh, in which the dominant coloured mineral is a purplish-brown pleochroic augite, often with hourglass structure[6]. Deep brown hornblende is also present, chiefly as a marginal intergrowth with the augite. Felspar and biotite are subordinate. Most of the olivine is converted into a yellow serpentine. Augite-picrites with typical pœcilitic structure occur in Shropshire[7].

Intrusions of *enstatite-picrite* occur in the old gneiss of the west of Sutherland. In one near Lochinver the slightly pleochroic enstatite or bronzite moulds the olivine, but shews good crystal-faces, being enclosed by large crystal-plates of felspar. There is a subordinate colourless augite and some brown hornblende, which is partly formed from the pyroxenes, partly original and later than the felspar. This rock is almost

[1] See Teall, Pl. VI.

[2] *Q. J. G. S.* xxxvii (1881) 137–140; xxxix (1883) 254–259. Also a similar rock from Alderney, *ibid.* (1889) xlv, 384.

[3] *G. M.* 1889, 109–112.

[4] *Amer. Journ. Sci.* (1886) xxxi, 31–37.

[5] Teall, Pl. IV, fig. 2, and Pl. VII.

[6] The augite resembles that common in nepheline-dolerites, and the rock differs in other respects from plutonic types.

[7] *Rep. Brit. Ass.* for 1887, 700; *Proc. Geol. Assoc.* (1894) xiii, 340, with figure.

as much a norite as a picrite, but true enstatite-peridotites also occur in the district, consisting of about equal parts of olivine and a rhombic pyroxene, with grains of pleonaste (fig. 19).

Of *mica-peridotite* few examples are described. One from Kentucky[1] consists of serpentinized olivine and pale yellow-brown to colourless mica, with pœcilitic arrangement, besides crystals of perofskite, *etc.* Another rock occurs in association with the gabbros of the Harz[2]. Prof. Judd[3] has described under the name 'scyelite' a *hornblende-mica-peridotite* from the borders of Sutherland and Caithness (Loch Scye and

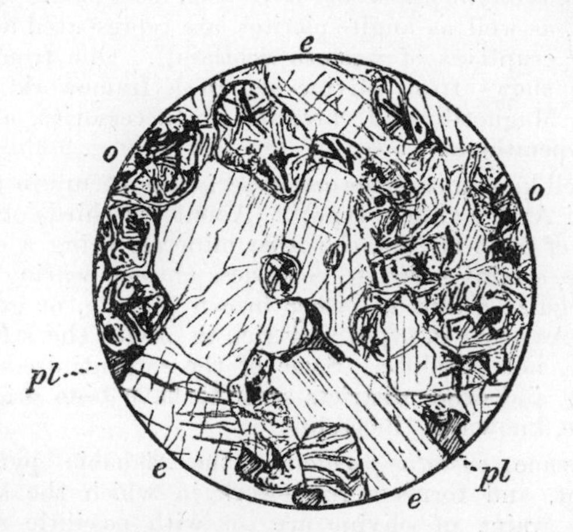

FIG. 19. ENSTATITE-PERIDOTITE, ASSYNT LODGE, SUTHERLAND; × 20.

A granular aggregate of olivine (*o*), largely serpentinized, and a slightly pleochroic enstatite or bronzite (*e*). These two minerals are in about equal quantity: in addition there are little irregular grains of isotropic green pleonaste (*pl*) [1642].

Achavarasdale Moor). Here serpentinized grains of olivine are enclosed in pœcilitic fashion by a pale green to colourless hornblende, probably pseudomorphous after diallage, and a

[1] Diller, *Amer. Journ. Sci.* (1892) xliv, 286–289.
[2] Koch, *M. M.* ix, 41, 42 (*Abstr.*).
[3] *Q. J. G. S.* (1885) xli, 401–407. Teall, Pl. v, fig. 2.

peculiar yellow mica. Dr Hatch[1] has noticed a handsome *hornblende-hypersthene-peridotite* from the neighbourhood of Kilimanjaro.

Among *hornblende-peridotites* we may place the rock described as a hornblende-picrite from Greystones in Wicklow[2], which is non-felspathic. The dominant hornblende is green, and encloses in pœcilitic fashion the olivine-pseudomorphs (of magnetite and a carbonate). It has cores and borders of colourless hornblende, and there is a third variety of this mineral with few cleavage-cracks and much magnetite dust.

Various *augite-peridotites* have been described. Specimens of these, as well as augite-picrites, are represented among the Tertiary eruptives of western Scotland[3]. One from the Isle of Rum shews fresh olivine set in a framework of green augite. Magnetite and chromite are accessories, and sometimes hypersthene.

A well-known *enstatite-augite-peridotite* occurs in the Pyrenees and Ariège (Lherz type)[4]. About two-thirds of the rock consists of fresh olivine, the other minerals being a colourless enstatite, a faint green to colourless chrome-bearing diopside, and irregular grains of either brown picotite, or green pleonaste. As usual in types very rich in olivine the structure is granular, not pœcilitic. Some of the serpentinized rocks of Cornwall and other districts seem to have been originally of this type, known as 'lherzolite.'

In some *enstatite-peridotites* the rhombic pyroxene is abundant, and forms a framework in which the somewhat rounded grains of olivine are set with pœcilitic structure. A well-known representative comes from the Harz (Baste or Harzburg type)[5], where, however, both minerals are more or less completely serpentinized.

In another type olivine largely predominates, and the enstatite occurs in relatively large crystals, which, among the

[1] *G. M.* 1888, 257–260.
[2] Watts, *Rep. Brit. Assoc.* for 1894.
[3] Judd, *Q. J. G. S.* (1885) xli, 389–395.
[4] Bonney, *G. M.* 1877, 59–64. For coloured figures, see Teall, Pl. I, fig. 1; Fouqué and Lévy, *Min. Micr.* Pl. LII, fig. 1.
[5] *Ibid.* Pl. LIII, fig. 2.

smaller grains of olivine, give a pseudo-porphyritic appearance to the rock. Good examples occur near Tromsö, *etc.*, in Norway[1] (fig. 18). In Maryland[2], Williams has described similar rocks in which large crystals of bronzite or diallage, or both, are imbedded in a granular mass, mainly of olivine.

From these rocks it is only a step to one composed wholly of olivine, with only a little accessory picotite or magnetite. Of this pure *olivine-rock* the type comes from New Zealand (Mount Dun), and is the 'dunite' of Hochstetter. Other examples might be named, *e.g.* from Kraubath in Styria, from St Paul's Rocks in mid-Atlantic[3], *etc.*, with others more or less serpentinized.

Of *garnet-peridotites* that from Elliott County, Kentucky[4], is a good example. The pyrope crystals are surrounded by a 'celyphite' border of brown mica with an outer ring of magnetite-dust, these minerals being supposed to be due to a reaction between the garnet and the olivine. The serpentine-rock of Zöblitz in Saxony is another example, in which, however, the olivine is wholly destroyed. Garnet occurs as an accessory in the diallage-peridotite of Tunaberg in Norway[5] (the 'eulysite' of Erdmann) and in other localities.

Serpentine-rocks. Hitherto we have noticed only very briefly the secondary changes that affect the minerals of crystalline rocks. In the present family, however, the decomposition of a rock is often so complete that its original nature is detected only by careful study, and the altered rock-masses are commonly denoted by a special name, serpentine-rocks or simply serpentines, expressing their dominant mineral composition. The mineral serpentine is the commonest decomposition-product of the non-aluminous magnesian silicates

[1] For an example from New Zealand, see Ulrich, *Q. J. G. S.* (1890) xlvi, 625–629, Pl. XXIV. This rock contains grains of nickel-iron-alloy (awaruite).

[2] *Bull. No.* 28 *U. S. Geol. Surv.* (1886) 50–55; *Amer. Geol.* (1890) vi, 38, 39, Pl. II, fig. 1.

[3] Renard, *Voyage of 'Challenger,'* Narrative, vol. ii, Append. B, with plate.

[4] Diller, *Amer. Journ. Sci.* (1886) xxxii, 121–125; *Bull.* 38 *of U. S. Geol. Surv.* (1887).

[5] Wadsworth, *Lith. Stud.* p. 147.

(olivine, the rhombic pyroxenes, and some of the augites and hornblendes), and the purest serpentine-rocks result from the alteration of peridotites[1]. Other decomposition-products occur in the rocks, *viz.* iron-oxides (magnetite and limonite), steatite, carbonates (dolomite, *etc.*), chlorite, and tremolite, but the bulk is serpentine of various kinds, in which may be found undestroyed relics of the original minerals of the peridotite (olivine, diopside, pyrope, chromite, *etc.*).

Of the mineral serpentine some kinds are crystalline and doubly refracting with the interference-colours of quartz or felspar, and faint pleochroism when the green tint is sufficiently pronounced. The habit is fibrous (chrysotile) or scaly (antigorite, *etc.*). Other kinds are amorphous and sensibly isotropic. Much of the serpentine occurs in definite pseudomorphs, and often retains something of the structure of the parent mineral to indicate its source. We may distinguish four cases:

(i) Serpentine derived from olivine, with the '*mesh-structure*[2]' (Tschermak's Maschenstructur; see p. 63 and fig. 20).

(ii) Serpentine derived from enstatite or bronzite, in distinct pseudomorphs with the *bastite-structure* (see p. 62 and fig. 20).

(iii) Serpentine derived from a non-aluminous hornblende, with '*lattice-structure*[3]' (Gitterstructur of Weigand). Here the cleavage of the hornblende is marked by veins of birefringent serpentine in two sets making the characteristic angle $55\frac{1}{2}°$. This serpentine is minutely fibrous, with the fibres set perpendicularly to the cleavage of the hornblende. The rest of the pseudomorph is of serpentine giving no definite crystalline reaction and consisting probably of a confusedly fibrous aggregate.

[1] For descriptions and coloured figures of numerous serpentine-rocks, see Wadsworth, *Lithological Studies* (1884). For a general sketch of observations and opinions on serpentine, see Teall, Chap. vi. On serpentine from diopside, see Merrill, *Proc. U. S. National Mus.* (1888) xi, 105–109, Pl. XXXII; *Am. Jl. Sci.* (1889) xxxvi, 189–191.

[2] Rosenbusch-Iddings, Pl. XXVI, fig. 4.

[3] *Ibid.* fig. 5.

(iv) Serpentine derived from a non-aluminous augite, with '*knitted structure*[1]' (gestrickte Structur of Hussak). This consists chiefly of serpentine with scaly habit (antigorite). The scales give straight extinction and low polarisation-tints. They occur in two closely interlacing sets parallel to the cleavage-planes of the augite, and so making an angle of about 87° with one another.

The source of serpentine in rocks can often be made out by these various characters, and it is placed beyond doubt when any unaltered remnants of the parent mineral remain. In addition there may be serpentine encroaching upon contiguous minerals or traversing them in veins: this is, as a rule, sensibly isotropic.

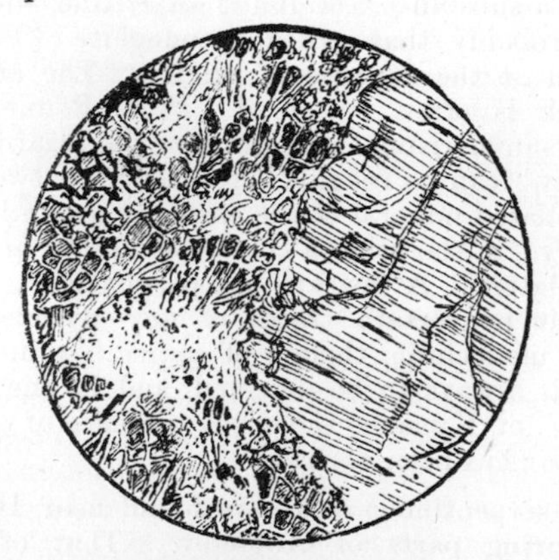

FIG. 20. SERPENTINE-ROCK, COVERACK, CORNWALL; ×20.

A large bastite-pseudomorph after bronzite is seen on the right. The rest of the rock is of serpentine with mesh-structure, derived from olivine: it is stained in places with hydrated iron-oxide [1118].

The best known serpentine-rocks in this country are those of the Lizard district in Cornwall[2]. The purer examples consist essentially of serpentine of various kinds, secondary

[1] Rosenbusch-Iddings, Pl. XXVI, fig. 6.

[2] Bonney, *Q. J. G. S.* (1877) xxxiii, 915–923; and (1883) xxxix, 21–23; Teall, 115 *et seqq.*

iron-ore (often peroxidised), steatite, tremolite, *etc.*, and often undestroyed relics of olivine or other characteristic minerals of the peridotites. Professor Bonney has shewn that much of the serpentine has the character of that derived from olivine, and some of the original rocks were probably nearly pure olivine-rocks (Dun type). Others were enstatite- or bronzite-peridotites, and shew large bastite-pseudomorphs after a rhombic pyroxene (Cadgwith, Coverack, *etc.*; fig. 20, cf. fig. 18)[1]. Others again are altered hornblende-peridotites, some of the serpentine shewing the mesh- and some the lattice-structure, while relics of olivine, hornblende, and picotite may remain (Mullion Cove, Kynance Cove, *etc.*)[2]. Augite-picrites are also represented (Menheniot, *etc.*). Here felspar has been altered into a substance resembling serpentine, which Mr Teall thinks is probably that called pseudophite. Tremolite has been formed at the expense of olivine. The augite of the original rock is often preserved. Prof. Bonney and Gen. McMahon[3], summarising the features of the Lizard serpentines, say that they "can be roughly separated into two groups: in the one a foliated mineral of the enstatite group is a conspicuous accessory; in the other a colourless augite or hornblende, usually the latter. A few are non-porphyritic[4], and in some cases exhibit no certain traces of any pyroxenic mineral, rhombic or monoclinic, though of course a spinellid or some iron oxide is always to be detected, and in one instance (at the Rill[5], W. of Kynance Cove) the presence of a fair proportion of felspar has been asserted."

Various serpentinous rocks are found near Holyhead and in neighbouring parts of Anglesey. That of Ty-ucha is regarded by Prof. Bonney[6] as an altered olivine-rock: in other examples there is much calcite-veining, producing 'ophicalcite' (Cruglas)[7]. In rocks at Four-mile Bridge much of the

[1] See also Teall, Pl. I, fig. 2.
[2] See Teall, Pl. XV.
[3] *Q. J. G. S.* (1891) xlvii, 466.
[4] In the sense of containing no conspicuous crystals.
[5] Teall, p. 119. "The original rock, therefore, was of the nature of a picrite." See also *G. M.* 1887, 137, 138.
[6] *Q. J. G. S.* (1881) xxxvii, 45.
[7] Blake, *Rep. Brit. Assoc.* for 1888, p. 409.

serpentine has the character of that derived from augite, and the parent-rock seems to have been genetically connected with a gabbro mass. Mr Blake, however, finds indications of olivine- and enstatite-serpentine[1].

Of the numerous serpentine-rocks of Scotland, one at Balhamie Hill in Ayrshire has been described by Prof. Bonney[2] as an altered olivine-bronzite-rock, closely resembling that of Cadgwith in Cornwall, the structure being of the pseudo-porphyritic type. Some near Belhelvie in Aberdeenshire[3] have also been enstatite-peridotites, but with the pœcilitic structure, and now shew pseudomorphs after olivine set in a framework of bastite, just as in the rock of Baste in the Harz[4], which has given its name to the latter mineral.

It is commonly believed that the mineral serpentine is in all cases a decomposition-product of other magnesian silicates. Recently, however, Weinschenk has maintained that it may occur under certain conditions as an original constituent of a peridotite. His Stubach type, from the Venediger district in the East-Central Alps, consists essentially of olivine and serpentine, intergrown in crystallographic relation, and he believes both minerals to be formed from igneous fusion.

1 Blake, *Rep. Brit. Assoc.* for 1888, p. 408.
2 *Q. J. G. S.* (1878) xxxiv, 770.
3 *G. M.* 1885, 439–448.
4 Fouqué and Lévy, Pl. LII, fig. 2.

B. INTRUSIVE ROCKS.

THE name 'intrusive,' which, in the absence of a better, is here applied to a large division of igneous rocks, is perhaps open to some objection. Plutonic masses are also in general intrusive on a large, and sometimes on a small scale, while, near volcanic vents, dykes occur of rocks practically identical petrologically with superficial lavas. The essential fact for our purpose, however, is that certain families of rocks are met with almost exclusively in the form of dykes, sills, laccolites of moderate dimensions, and 'pipes' of old volcanoes; and that the rock-types comprised in these families have characters differing from those of both the plutonic and volcanic divisions, and in many respects holding an intermediate position between the two. They correspond in a general way, though not precisely, with the 'dyke-rocks' (Ganggesteine) of Rosenbusch.

It must be admitted that, from the descriptive point of view, the rocks here included do not form a natural group with a well-defined set of characters in common distinguishing them from the other groups. Most of them are holocrystalline, but in some a glassy residue is found. In some families the porphyritic structure is characteristic, as it is in the volcanic rocks; in others it is wanting or non-significant: but even the holocrystalline non-porphyritic types have structural and mineralogical characters, to be noted below, which differentiate them from rocks of truly deep-seated origin.

CHAPTER VII.

ACID INTRUSIVES.

THE acid intrusive rocks embrace a considerable range of varieties, bridging over the difference between the even-grained, holocrystalline granites and the porphyritic, largely glassy rhyolites. The porphyritic character is almost universal, but the ground-mass which encloses the phenocrysts may be holocrystalline, partly crystalline and partly glassy, or wholly glassy. On the nature and special structures of the ground-mass depend the several types usually recognized among these rocks. All agree in that the constituent minerals—in so far as these are developed—include in the first rank felspars rich in alkali and usually quartz, while ferro-magnesian minerals and free iron-ores occur only in relatively small quantity, and are sometimes wanting.

From an examination of their mineral constitution and characteristic structures, the more crystalline types are readily referred to their proper positions; but, in proportion as the bulk of the rock comes to consist of unindividualised glassy matter or an irresolvable cryptocrystalline 'base,' the criteria become fewer. In particular, the first stage of consolidation (that of the phenocrysts) may have been arrested before quartz (the last mineral) began to crystallize, and so, if the ground-mass consolidates as a glass, we may have a thoroughly acid rock without quartz. Thus the most glassy rocks (pitchstones) belonging to this family are not always to be distinguished by the microscope alone from less acid pitchstones. Again, they are scarcely divided from some glassy rhyolites (obsidians).

The nomenclature of the acid intrusives is confused. The name 'felsite' or—if containing evident phenocrysts of quartz—'quartz-felsite' has been applied in this country not only to these rocks but also to many volcanic rocks (acid and intermediate); and their usage lacks precision and significance. The name quartz-porphyry, borrowed from the German, covers most of the rocks, but not all, since porphyritic quartz may be wanting; this term is also used by Continental writers for the 'older' acid lavas. For a type rich in soda, and having some mineralogical peculiarities, the name quartz-ceratophyre (Ger. Quarz-Keratophyr) has been used. It will be convenient to speak of the family, as a whole, as the acid intrusives. The names applied to particular types will be noticed in connection with the ground-mass.

Constituent minerals. We notice here especially the minerals occurring as phenocrysts. Of these, the felspars include *orthoclase* (not microcline) and an acid plagioclase such as *oligoclase.* The two are commonly associated, and both build idiomorphic crystals with the usual types of twinning. A narrow zone of orthoclase surrounding each plagioclase crystal is seen in some rocks. The characteristic felspar of the ceratophyres is *anorthoclase.*

The *quartz* has crystallized in the ordinary hexagonal pyramids, sometimes with narrow prism-faces, but the crystals are frequently rounded and eaten into, owing to corrosion by the ground-mass, and may have lost all crystal outlines. In the rock-types most nearly approaching granites (granite-porphyries) the quartz contains fluid-pores: in other types the inclusions are mostly of glass or portions of the ground-mass. As already mentioned, quartz-phenocrysts are not always present.

The brown *biotite*, which occurs in many of the rocks, has the same characters as in granites, and carries the same inclusions. It is usually in good hexagonal flakes. Less commonly, in the marginal part of an intrusion, it has a blade-like habit, due to extension along the *a*-axis. Hexagonal flakes of *muscovite* are found in a few of the granite-porphyries only.

A green *hornblende* in well-built crystals is a rather exceptional constituent. The deep blue soda-bearing amphibole *riebeckite* occurs in a few rocks, but always in very ragged allotriomorphic crystals. The *augite* of the acid intrusives is a pale greenish variety like that in some granites, but occurs here much more frequently. It builds good idiomorphic crystals in many granophyres and pitchstones. A rhombic pyroxene is of rarer occurrence.

As accessories, *apatite* and *zircon* are widely but sparingly distributed, while the iron-ores are usually represented only by a little *magnetite*. Such minerals as garnet, tourmaline, and pinite pseudomorphs after *cordierite*[1] occur in special localities.

Ground-mass and structures. The types which approach most nearly to the plutonic habit are known as *granite-porphyry*. Here relatively large idiomorphic crystals of quartz and felspars, with mica or some other ferro-magnesian mineral, are enclosed in a fine-textured crystalline ground-mass of felspar and quartz. The structure of this ground may resemble that of a granite, or may be distinguished by a more marked idiomorphism of the lath-shaped felspars, usually untwinned. Mica may also occur in a second generation as part of the ground-mass.

Very common are the types in which the phenocrysts, consisting of felspars, more or less corroded quartz, and biotite or some other constituent, are imbedded in a very finely crystalline ground-mass of felspar and quartz. The elements of the ground-mass may have more or less idiomorphism. Quartz-porphyries having an evidently microcrystalline ground-mass of this kind are styled by Rosenbusch *microgranites*, the porphyritic character being understood.

When the texture of the ground-mass sinks to such minuteness as to be not clearly resolved under the microscope, it may be described as *cryptocrystalline* ('microfelsitic' of some authors). For such rocks Rosenbusch uses the term felsophyre[2]. Without entering into a discussion of an obscure

[1] Fouqué and Michel Lévy, *Min. Micr.* Pl. XIII, fig. 5.
[2] *Cf.* Teall, *G. M.* 1885, 108–111.

subject, it may be said that this cryptocrystalline ground is probably in some cases original, in other cases due to secondary change (devitrification) of a ground-mass originally glassy.

The glassy (or 'vitrophyric') type of ground-mass is seen in the rocks known as *pitchstones.* In some of these, phenocrysts of felspar, *etc.*, are only sparingly present, the great bulk of the rock consisting essentially of isotropic glass. This glassy 'base,' however, includes in many cases innumerable minute and imperfectly developed crystalline growths (*crystallites*) with regular grouping (fig. 23). These minute bodies will be more fully noticed in connection with the acid lavas. The pitchstones frequently shew perlitic cracks, and occasionally some of the flow-phenomena, which are better exhibited in lavas. Typical pitchstones, excluding lava-flows, are of quite limited distribution.

In the above types we have what may be regarded as a graduated transition from the granitic to the rhyolitic structures, the only gap, that between cryptocrystalline matter and glass, being one which the instruments at our disposal do not enable us to bridge. There is, however, a second, more or less distinct, line of transition, parallel to the former but characterized by a different set of structures, *viz.* micrographic intergrowths of felspar and quartz and regular crystalline aggregates of felspar fibres. To these structures Rosenbusch applies the somewhat inappropriate term 'granophyric,' including both micropegmatitic and microspherulitic: and the rocks having a ground-mass of this nature are very generally known as *granophyres.*

We have already noticed in some granites a micrographic intergrowth of the kind named micropegmatite; but when the whole mass of the rock, exclusive of crystals of certain minerals, takes on this character, we have a type characteristic of intrusive rather than plutonic rocks as here understood. In such rocks the quartz and the greater part of the felspar form a micrographic ground-mass, which may enclose idiomorphic crystals of some ferro-magnesian mineral (augite or biotite) or of felspar (mostly plagioclase). Further, the micrographic intergrowth may come in to some extent in rocks which on the whole would be placed with the granite-porphyries or the microgranitic type. When the intergrowth is on a relatively coarse scale, it is often rude and irregular, but the finer-textured '*micropegmatite*' shews great regularity and

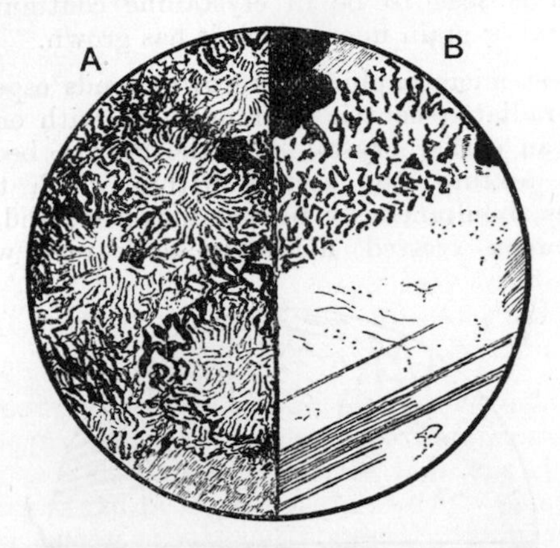

FIG. 21. GRANOPHYRES, SHEWING MICROGRAPHIC INTERGROWTH OF FELSPAR AND QUARTZ ; ×20.

Crossed nicols. *A*. Crug, near Caernarvon: shewing an intricate aggregate of rather delicate micropegmatite with a tendency to irregular 'centric' arrangement [17]. *B*. Carrock Fell, Cumberland, shewing part of a phenocryst of oligoclase with a fringe of micropegmatite. The felspar in this is in crystalline continuity with the phenocryst; the quartz, shewn in the position of extinction, is continuous with a quartz-grain at the top of the figure [1545].

often a definite arrangement (Fig. 21, *A*). In particular it frequently forms a regular frame surrounding phenocrysts of felspar[1], and it can often be verified that the felspar of the intergrowth is in crystalline continuity with the felspar crystal which served as a nucleus (Fig. 21, *B*). The appearance is as if the original crystal had continued to grow throughout the final consolidation of the rock, enclosing the residual excess of silica as intergrown quartz. Sometimes a

[1] For good illustrations see Irving, *Copper-bearing Rocks L. Superior*, Pl. XIV, figs. 1, 2.

line of Carlsbad twinning can be traced from the crystal through the surrounding frame. There is no doubt that plagioclase felspar, as well as orthoclase, enters into such micrographic intergrowths. Less frequently the quartz of the intergrowth is seen to be in crystalline continuity with a quartz crystal or grain upon which it has grown.

The finest micrographic intergrowth tends especially to a stellate or radiate ('centric') arrangement, with or without a nucleus of an earlier crystal. As the growth becomes very delicate in texture, the sectors within which the felspar extinguishes simultaneously become narrower, and are represented between crossed nicols by dark rays when their

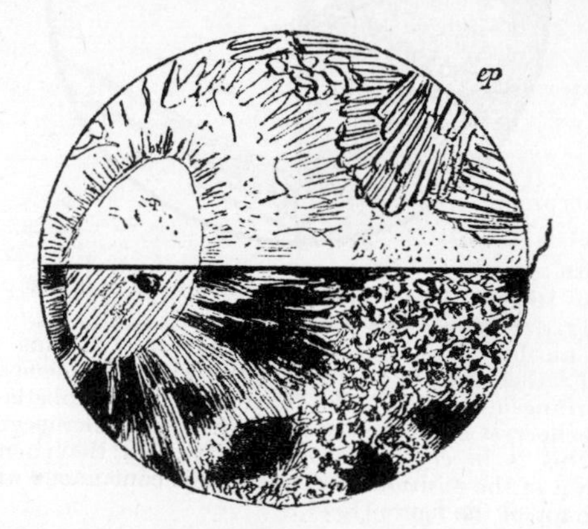

FIG. 22. GRANOPHYRE (SPHERULITIC QUARTZ-PORPHYRY), ST DAVID'S; ×20.

Upper half in natural light, lower half between crossed nicols. A cryptographic intergrowth (pseudospherulitic of some authors) is grown round a corroded quartz-grain. The bundle of highly refracting crystals (*ep*) is secondary epidote [350].

direction makes a small angle with one of the cross-wires. When the structure is on too minute a scale to be resolved by the microscope, it may be termed, by analogy, cryptographic. (Fig. 22). The optical characters of such an aggregate appear to be determined by the minute radially arranged fibres of felspar, which obscure the quartz. The structures known as *microspherulitic* and pseudo-spherulitic in acid rocks are probably of this nature. Between crossed nicols they shew characteristically a black cross, caused by extinction in those fibres which lie nearly parallel to one of the cross-wires. Such growths cluster round porphyritic crystals of quartz or felspar, or, as innumerable closely packed minute spherules, constitute almost the whole of the ground-mass[1].

Isolated spherulites or bands of spherulites may occur in a vitreous ground.

British examples. This country affords examples of acid intrusive rocks in great variety, and it will be sufficient to note a few of these, illustrating the several points indicated above. In view of the frequent association of the different types of ground-mass in one district or even in parts of one intrusion, we shall not find it convenient to follow any strict order.

An intrusion near Dufton Pike[2] in Westmorland is a characteristic granite-porphyry with both white and dark micas, which occur both as phenocrysts and in the ground-mass. The other phenocrysts are idiomorphic quartz and felspar, chiefly plagioclase but with a few large sanidine-crystals. A marginal modification of the rock shews the blade-like habit of the biotite.

The Carboniferous 'elvan' dykes of Cornwall and Devon, as described by Mr J. A. Phillips[3] and by Mr Teall, have a microcrystalline to cryptocrystalline ground-mass enclosing large felspars, pyramidal or rounded quartz crystals, and often mica. Tourmaline is of frequent occurrence in crystals or stellate groups of needles, and is sometimes seen to replace felspar. An occasional constituent is cordierite, represented by the so-called 'pinite' pseudomorphs of yellowish green micaceous flakes (Sydney Cove[4]).

[1] For good figures of micrographic and cryptographic structures, ranging from the micropegmatitic to the spherulitic, see Fouqué and Michel Lévy, *Min. Micr.*, Plates x, fig. 2, xi, fig. 1, xii, xiv, xv, xvi.

[2] *Q. J. G. S.* (1891) xlvii, 519.

[3] *Q. J. G. S.* (1875) xxxi, 334–338, Pl. xvi.

[4] Teall, p. 334.

The varied group of Ordovician intrusive rocks in Caernarvonshire[1] include some granite-porphyries of a well-marked type. Quartz is wanting among the phenocrysts, which are chiefly of oligoclase. One example at the head of Nant Ffrancon has a ground-mass of allotriomorphic quartz and felspar (chiefly orthoclase). The ferro-magnesian constituent is biotite. Others, quarried at Yr Eifl and near Nevin, have a ground in which idiomorphic felspars are moulded by interstitial quartz. These contain augite, usually without biotite[2]. Other rocks in the district, all augitic, shew more or less tendency to micrographic structures, and in many the whole ground-mass is of micropegmatite. Beautiful examples occur in the hills above Aber and at Moel Perfedd in Nant Ffrancon. The growth of the micropegmatite round felspar crystals is well exhibited, and in some cases a narrow zone of orthoclase is seen interposed between a plagioclase crystal and the surrounding growth. The structure is rarely so minute as to approximate to the spherulitic. Many of the smaller intrusions in the district, *e.g.* near Clynog-fawr, are of quartz-porphyry with a cryptocrystalline ground, which may possibly be due to devitrification. Porphyritic quartz, which is wanting in the more evidently crystalline types, appears here in corroded crystal-grains. A somewhat similar rock is that forming a low range in the neighbourhood of Llanberis. This exhibits flow-structure in places, and has been considered by Professor Bonney and others as a group of lavas.

The complex group of acid rocks near Caernarvon and eastward, which some have supposed to be of pre-Cambrian age, afford examples of granite-porphyries, micrographic rocks (Fig. 21, *A*), microcrystalline and spherulitic quartz-porphyries, *etc.* The spherulitic growths often surround pyramids of quartz. The porphyritic felspars in all these rocks are mostly plagioclase, and the ferro-magnesian mineral is biotite, often green from alteration. Various granophyres and, especially, beautiful spherulitic rocks, shewing the growth round pyramidal crystals

[1] *Bala Volc. Ser. Caern.* 48–56.

[2] In the rock of Mynydd Mawr near Nantlle occurs the blue amphibole, riebeckite: *ibid.* 50. *G.M.* 1888, 221, 455.

of quartz, occur at St David's[1]. The structure is of the cryptographic type, not shewing a very perfect black cross (fig. 22).

The Lake District contains examples of microgranites, such as the rock quarried at Threlkeld, while some minor intrusions shew a cryptocrystalline ground. Granophyres also occur, the large Buttermere and Ennerdale intrusion being of a micropegmatitic rock with either biotite or augite, resembling some Caernarvonshire examples. The dykes of Armboth and Helvellyn have a spherulitic ground-mass enclosing idiomorphic crystals of quartz and felspar. The spherulitic growth, which does not always give a good black cross, is clustered especially about the quartz crystals. A few garnets occur. These rocks are probably all Ordovician. The Devonian dykes about Shap, in Edenside, near Sedbergh, *etc.*, have microcrystalline to cryptocrystalline grounds, and some of them contain biotite rather abundantly.

One of the most beautiful granophyres in this country is that of Carrock Fell, in Cumberland[2]. It contains a pale augite in good crystals, often uralitised or otherwise altered, and rarely a little biotite. There are also idiomorphic felspars, usually oligoclase, and some granules of iron-ore. The ground-mass shews in different specimens, or even in one slide, every gradation, from a coarse irregular micropegmatite through exquisitely regular micrographic[3] and cryptographic structures to what would be described as spherulitic. These intergrowths usually make up the whole ground-mass, though sometimes part of the quartz forms irregular grains. The arrangement is sometimes centric, but more usually peripheral to the felspar phenocrysts, forming a regular border to them. It can often be seen that the felspar of the intergrowth is continuous with that of the crystal, and much of it must be plagioclase (Fig. 21, *B*).

Other augite-granophyres are found among the Tertiary intrusions of Scotland and Ireland, *e.g.* in Mull[4] and in the Carlingford district. Some are quite coarse micropegmatites,

[1] Geikie, *Q. J. G. S.* (1883) xxxix, 315, Pl. x, figs. 8, 9.
[2] Harker, *Q. J. G. S.* (1895) vol. l.
[3] Teall, Pl. XLVII, fig. 5 (misplaced 4 in key-plate).
[4] Teall, Pl. XXXIII, fig. 1.

or shew only a rude kind of intergrowth, and these rocks are frequently miarolitic. The more delicate micrographic and cryptographic growths are, however, also represented. Some of the Skye granophyres have riebeckite instead of augite[1]. A remarkable rock from Corriegills in Arran[2] appears as if divided into polygonal areas each enclosing a spherule with well-marked boundary and radial structure. Dr Hyland[3] has described granophyre dykes in Co. Down. These contain apparently no augite, but a little green hornblende (Newcastle) or brown mica (Hilltown).

The granophyre of Stanner Rock near New Radnor has perhaps been augitic, but the ferro-magnesian mineral is now a fibrous hornblende. From Prof. Cole's description[4] the rock seems to be of the cryptographic type. The biotite-bearing quartz-porphyries of the Cheviots[5] have sometimes granophyric structures, but are more commonly micro- or cryptocrystalline. Mr Kynaston has shewn me numerous examples in which the ground-mass encloses patches of micropegmatite like porphyritic crystals, sometimes shewing the outlines of idiomorphic felspar.

Hornblende-granophyre occurs in the Grampians. A specimen from Beinn Alder contains phenocrysts of oligoclase, twinned green hornblende, biotite, and magnetite in a groundmass of delicate micropegmatite.

The finest examples of pitchstones are those of Arran[6], of which some are of acid, others of subacid composition. They form dykes, probably of Tertiary age. The phenocrysts are of sanidine, quartz, plagioclase, and augite, varying in different examples and sometimes occurring very sparingly. The ground-mass is of glass crowded with crystallites, which often

[1] The same is true of the microgranite of Ailsa Craig; Teall, *M. M.* (1891) ix, 219–221; *Q. J. G. S.* (1894) 219.

[2] Allport, *G. M.* 1872, 541; Bonney, *G. M.* 1877, 506–508.

[3] *Sci. Proc. Roy. Dubl. Soc.* (1890) vi, 420–430.

[4] *G. M.* 1886, 220–222, with figs.

[5] Teall, *G. M.* 1885, 111.

[6] Allport, *G. M.* 1872, 1–9; 1881, 438: Bonney, *G. M.* 1877, 499–511: Judd, *Q. J. G. S.* (1893) xlix, 546–551, 559–561, Pl. XIX: Geikie, p. 116, fig. 14: Teall, Pl. XXXIV, figs. 3, 4. Cf. Sollas on Donegal pitchstones, *Sci. Proc. Roy. Dubl. Soc.* viii, 87–91 (1893).

assume peculiar groupings. In one variety needle-shaped microlites (belonites) of hornblende occur, each forming the trunk of a delicate arborescent aggregate of more minute bodies (Corriegills, fig. 23, *A* and *B*). In another variety occur crosses, each of the four arms carrying a plume-like

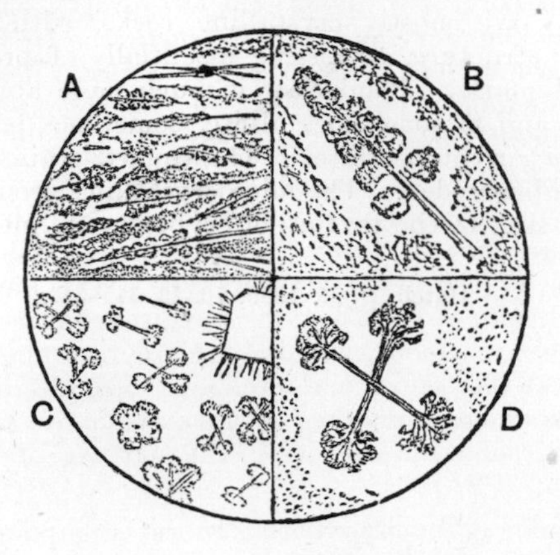

FIG. 23. PITCHSTONES, ARRAN.

A. Arborescent grouping of crystallites, Corriegills; ×20. *B*. The same, ×100 [57]. *C*. Plumed cross-like groupings, and growth of crystallites on a small felspar phenocryst, Tormore; ×20. *D*. The same, ×100 [G. 73].

growth (Tormore, fig. 23, *C* and *D*). Again, little rod-like bodies frequently occur as a fringe arranged perpendicularly on the faces of phenocrysts. The general mass of the glass is full of very minute crystallitic bodies, but around each grouping is a clear space, indicating that the tree-like or other growth has been built up at the expense of the surrounding part. Flow-structures are only occasionally met with, and perlitic cracks are not common. The latter are well shewn by 'pitchstones' from the Is. of Eigg[1]. These are rocks rather rich in phenocrysts of sanidine, and having the glassy base rich in crystallitic growths but without arborescent grouping.

[1] Teall, Pl. XXXIV, fig. 5 [454].

They belong, however, like the 'Meissen pitchstones' of Saxony, to the rhyolite family, being extruded lava-flows.

Acid intrusives rich in soda (quartz-ceratophyres) are not yet well known in this country. Probably some of the 'soda-felsites' of Leinster[1], of Ordovician age, are to be placcd here. They are mostly crystalline rocks, with or without porphyritic structure, consisting essentially of predominating felspar and quartz. Plagioclase is much more abundant than orthoclase, and is sometimes albite, sometimes possibly anorthoclase or cryptoperthite. The quartz-grains are often rounded and corroded. The microcrystalline ground does not apparently sink to the texture of cryptocrystalline.

[1] Hatch, *G. M.* 1889, 70–73, 545–549.

CHAPTER VIII.

PORPHYRIES AND PORPHYRITES.

THE rocks which are for convenience grouped together in this chapter belong to various intrusive types of intermediate chemical composition. They have not a very wide distribution, and they graduate on the one hand into the acid intrusives already discussed, on the other into the more peculiar family of the lamprophyres.

The porphyritic structure characterises all the rocks in question, and in most of the types is marked by felspar phenocrysts of relatively large size. The ferro-magnesian minerals are often confined to the elements of the earlier period of crystallization. Original quartz is found in the more acid types only, and is almost always restricted to the ground-mass.

The rocks may be regarded as standing between the plutonic syenites, diorites, *etc.*, on the one hand, and the volcanic trachytes, dacites, and andesites on the other, just as the rocks treated in the preceding chapter stand between the granites and the rhyolites. According as the dominant constituent is an alkali-felspar or a soda-lime-felspar, they fall into two families, to be distinguished as porphyries and porphyrites respectively.

Under the former head we may recognise *syenite-porphyry* and *orthoclase-porphyry* (or simply porphyry), corresponding to granite-porphyry and quartz-porphyry among the acid rocks. From these orthoclase-bearing rocks have been separated others characterized by a potash-soda-felspar, under the name *cerato-phyre* (Ger. Keratophyr). There are also nepheline-syenite-porphyry and nepheline-porphyry (tinguaite, *etc.*), which are of very restricted occurrence.

Of the rocks characterized by soda-lime-felspars, the types most nearly approaching the plutonic have been styled *diorite-porphyrite, etc.*, the others being termed simply porphyrites. Since some ferro-magnesian mineral is usually a prominent constituent, we have the divisions *mica-porphyrite, hornblende-porphyrite,* and *augite-porphyrite.* If a little porphyritic quartz be present, we have a *quartz-porphyrite* (quartz-mica-porphyrite).

It must be noted that writers who make no distinction in nomenclature between intrusive and volcanic rock-types use some of the above names in a more extended sense. Thus the Continental petrologists include under the term porphyrite the 'older' andesitic lavas, while some British authors apply the same name to andesites modified by secondary changes (partial decomposition, *etc.*). Some of the rocks styled propylites belong to the division now to be considered.

Constituent minerals. The *orthoclase* phenocrysts of the porphyries are similar to those in the quartz-porphyries and other acid intrusives. In the porphyrites this mineral does not occur except in the ground-mass. A *plagioclase* felspar accompanies the porphyritic orthoclase in some of the porphyries, and forms the most conspicuous phenocrysts in the porphyrites. Here it builds idiomorphic or rather rounded crystals, with twinning often on two or three different laws. It ranges in the porphyrites from oligoclase to labradorite, and frequently shews strong zoning between crossed nicols. A parallel intergrowth of orthoclase and plagioclase is common in some porphyries. In certain types of that family also occurs a felspar which has been referred to *anorthoclase,* while it has also been explained as a minute parallel intergrowth of a potash- and a soda-lime-felspar. As seen between crossed nicols a crystal is often seen to be divided rather irregularly into portions with different optical behaviour, sometimes one part finely striated, another without visible striation. In certain special rocks (rhomb-porphyries) the crystal has a peculiar habit, which gives a lozenge-shaped section; in the ceratophyres it has the usual habit, giving rectangular sections.

As phenocrysts *quartz* is found only sparingly in a few rocks, but it enters into the ground-mass of all the more acid of the porphyries and porphyrites, though less abundantly than in the true acid rocks.

The most usual ferro-magnesian minerals are brown *biotite* and a pale or colourless idiomorphic *augite*. Some of the porphyrites have *hornblende* in sharply idiomorphic prisms, often twinned: it is more usually brown than green.

As accessories, *apatite* and *iron-ores* (often titaniferous) may occur in varying quantity, the latter not being abundant. Exceptionally *olivine* and other minerals are present.

In the few rocks which contain *nepheline* or elæolite that mineral occurs in one or two generations. As phenocrysts it is idiomorphic, while the little crystals in the ground-mass may or may not have definite shape. The 'liebenerite' pseudomorphs in certain porphyries have been supposed to represent nepheline. They consist essentially of a pale mica, and may with equal probability come from the destruction of cordierite.

Ground-mass and Structures. In the great majority of the rocks here considered the ground-mass is holocrystalline, with a fine texture and with various types of structure. It consists essentially of felspar or, in the more acid members, of felspar and quartz. In the porphyries the felspar is usually in minute prisms, short in comparison with their length, and as a rule untwinned. Quartz, if present, occurs interstitially. The little prisms may have more or less of a parallel arrangement, due to flow. Such short and relatively stout prisms are usually referred to orthoclase: if the crystals have the 'lath'-shape, they are probably of a plagioclastic variety. Any approach to an allotriomorphic character is uncommon, and the micrographic intergrowths so frequent among the acid intrusives are not found here. In the nepheline-bearing rocks a more allotriomorphic type of structure is found.

The ground-mass of the porphyrites is also in general holocrystalline, consisting essentially of felspar or, in the most acid varieties, of felspar and quartz. In this latter case the rocks may reproduce some of the characteristic structures noted in the preceding chapter, such as the cryptocrystalline and the micrographic. Other porphyrites have the 'orthophyric' type of ground-mass (with short felspar-prisms), as in the porphyries, but there is every gradation from this to the allotriomorphic. In some of the more basic members the ground-mass consists of little lath-shaped plagioclase prisms with more or less noticeable flow-arrangement, an approach to the character of some andesites ('pilotaxitic' structure).

Glassy and vitrophyric rocks are not unknown in the families in question. Some of the Arran pitchstones, for example, have the composition of intermediate rather than acid rocks.

Illustrative examples. We shall select only a few examples to illustrate the characters of these rocks, and in some cases we must go for these examples to foreign localities.

A good instance of a *syenite-porphyry* is the rock quarried at Enderby, in Leicestershire. It contains phenocrysts of a strongly zoned plagioclase felspar and of pale greenish brown hornblende, with more sparingly flakes of biotite and round grains of quartz, in a moderately fine-textured ground-mass of quartz and felspar, apparently orthoclase.

The most usual type of *orthoclase-porphyry* (orthophyre of Rosenbusch) is exemplified by dykes and sills in the Carboniferous of Thuringia, in the Vosges, and in other districts. Besides the orthoclase phenocrysts there may be some of plagioclase. The ferro-magnesian minerals are only sparingly represented, and may be biotite, hornblende, or augite. The ground-mass is holocrystalline with the structure styled orthophyric, in which short prisms of untwinned felspar are associated with some interstitial quartz.

A peculiar type of restricted occurrence is that known as *bostonite*. It is recorded near Boston (Massachusetts) and in the Adirondacks, in Brazil, in the Christiania district, *etc.*, as dykes in connection with nepheline-syenite or other plutonic rocks, but especially in intimate association with dykes of lamprophyre (camptonite). The bostonites consist essentially of felspar, quartz being never abundant and the ferro-magnesian silicates typically absent. Phenocrysts may or may not be developed, the bulk of the rock being a ground-mass of little felspar rods, often with partial flow-disposition and recalling the structure of the trachytes. In some examples a high percentage of soda, with little or no plagioclase evident, points to a soda-orthoclase or anorthoclase, and indicates an affinity with the ceratophyres[1].

Among the Devonian intrusions of the Christiania district occur the singular rocks known as *rhomb-porphyry* (Ger. Rhombenporphyr), and they may be studied in numerous boulders in Holderness and the Eastern Counties. The phenocrysts of potash-soda-felspar, with their unusual crystallographic development, have been alluded to above. The crystals are often rounded and corroded, and they contain numerous inclusions of materials like the ground-mass. Some of the rocks contain pseudomorphs after olivine. The fine-textured holocrystalline ground-mass consists of short prisms of felspar (probably orthoclase) with little granules of augite. Apatite is often plentiful, and grains of titaniferous iron-ore occur.

The name *ceratophyre* was first used by von Gümbel for a rather varied group of rocks in the Fichtelgebirge. Somewhat similar rocks have been described from Saxony, Westphalia, the Harz, and other areas. Porphyritic quartz does not occur in the ceratophyres proper, and felspar is the predominant mineral in both phenocrysts and ground-mass. The phenocrysts have the peculiarities attributed to anorthoclase or to a cryptoperthite intergrowth. The commonest ferro-magnesian element is a pale augite (diopside). The felspar prisms of the ground-mass may be short and unstriated or lath-shaped and striated, and the more acid members have a little interstitial quartz.

[1] Kemp and Marsters, *Trans. N. Y. Acad. Sci.* xi (1891) 14–16.

Rosenbusch has given the name *tinguaite* to certain dyke-rocks which have the composition of the (plutonic) nepheline-syenites and the (volcanic) phonolites with structural characters which place them between those two families. Such rocks are associated with nepheline-syenites in the Serra do Tingua and other places in Brazil, in southern Portugal, in Arkansas, *etc.* Phenocrysts of orthoclase, often with marked tabular habit and with the characters of sanidine, are imbedded in a fine-textured holocrystalline ground-mass of orthoclase with elæolite or nepheline, ægirine, *etc.* This ground is typically allotriomorphic: when the little felspars take on the lath-shape with fluxional arrangement, the rocks do not differ essentially from phonolites. There may be phenocrysts of nepheline, and in one type (leucite-tinguaite) large pseudomorphs of orthoclase and elæolite occur in the form of leucite[1].

Among the *diorite-porphyrites*, the more acid varieties may contain a little quartz among the phenocrysts as well as in the ground-mass. Examples occur in association with the Banat type of quartz-diorite in Hungary (Szaska). Large, strongly zoned crystals of plagioclase, roughly idiomorphic green hornblende, and flakes of biotite, with magnetite, apatite, and some composite grains of quartz, are here enclosed in a microcrystalline ground of allotriomorphic felspar and quartz.

Porphyritic diorites or diorite-porphyrites free from quartz differ only in the respect indicated from normal diorites. The rock of Lac d'Aydat may be taken as an example.

Numerous *mica-porphyrite* dykes, of Old Red Sandstone age, occur in the Cheviots. The felspar phenocrysts (oligoclase-andesine) are frequently rounded, and shew carlsbad and albite twinning. The biotite-flakes are often bent, and sometimes shew a resorption border. A colourless augite may also occur, and magnetite and apatite are minor constituents. The ground-mass is microcrystalline, fine-textured, and often obscured by decomposition. Quartz plays a variable part in

[1] J. F. Williams, *Ann. Rep. Geol. Surv. Arkansas* for 1890, vol. ii, *The Igneous Rocks*, pp. 281–286. Compare O. A. Derby, *Q. J. G. S.* (1891) xlvii, 251–265.

it, and there are some transitions to granophyre and quartz-porphyry. Indeed the mica-porphyrites in general often carry a notable amount of quartz in their ground-mass.

The rock which forms large intrusive sills in the Torridon Sandstone of Canisp, Sutherland, may be placed here. It has large, frequently broken, phenocrysts of oligoclase, with carlsbad, albite-, and pericline-twinning. The dominant coloured mineral is biotite, but Mr Teall also notes augite, either colourless or green or the former bordered by the latter. Calcite pseudomorphs in the form of augite are common. These minerals, with some magnetite, are set in a fine microcrystalline ground-mass of felspar and quartz.

Among the rocks conveniently styled *hornblende-porphyrite* is the famous 'red porphyry' of the ancients (porfido rosso antico, porphyre rouge antique) obtained from Djebel Dokhan in Egypt. This contains good crystals of brown hornblende and of felspar (oligoclase-andesine) in a ground-mass essentially of little felspar prisms (apparently a more acid oligoclase). The minerals are considerably altered, and the chief decomposition-product, to which the red colour of the rock is due, is the manganese-bearing epidote, withamite, with vivid pleochroism, from bright rose-red to yellow-green.

For British examples we may take some of the rocks which form sills of Lower Palæozoic age in the Assynt district of Sutherland (Inchnadamff, *etc.*)[1]. Here the hornblende is green and in very perfect crystals, often twinned: they sometimes shew zonary colouring, and are occasionally hollow. A colourless augite in imperfect crystals sometimes accompanies the hornblende. The plagioclase phenocrysts shew strong zonary banding between crossed nicols. Magnetite and apatite are present sparingly. The microcrystalline ground-mass is of felspar with subordinate quartz. These rocks are part of a variable set of intrusions. On the one hand is a non-porphyritic and coarser textured type with allotriomorphic felspar (diorite), on the other a type with more abundant hornblende in two generations and with a panidiomorphic ground-mass (camptonite, see Chap. X. and fig. 29).

[1] Teall, *G. M.* 1886, 346–350.

A hornblende-porphyrite of basic composition is seen in the Mawddach valley, near Dolgelly. It contains large and rather irregularly bounded twin-crystals of brown hornblende in a much decomposed matrix. Mr Phillips[1] termed this hornblende uralite, but there is no clear evidence that it is other than an original mineral.

The rocks to which the name *augite-porphyrite* has been applied by German petrologists seem to be for the most part old augitic lavas, though intrusive types are also included. Such rocks, probably of Triassic age, are represented in the Monzoni district in the southern Tirol. The name *uralite-porphyrite* is correctly applied to augite-porphyrites in which the augite has been transformed into uralitic hornblende.

Types in which ferro-magnesian minerals are poorly represented, and the only noteworthy phenocrysts are those of plagioclase felspar, have been called by such names as *labradorite-porphyrite, etc.*, according to the nature of the porphyritic felspar. German writers have also applied the name diabase-porphyrite to such rocks, as well as to some of the 'older' andesitic lavas. As an example we may take the rock of Lambay, an island near Dublin, which forms an intrusion probably of Ordovician age. The porphyritic felspars are of a basic oligoclase, and are much decomposed, as is the ground-mass, which consists of plagioclase laths and partly destroyed augite, with much chloritoid material and some epidote, sphene, magnetite, pyrites, and calcite.

[1] *Q. J. G. S.* (1877) xxxiii, 427–429, Pl. xix.

CHAPTER IX.

DIABASES.

THE larger intrusive bodies of pyroxenic rocks, whether intermediate or basic in composition, have petrographical features which characterize them as a group with considerable individuality. It is to these rocks that we shall apply the name diabase. Like their plutonic equivalents, the gabbros, they are holocrystalline and typically non-porphyritic, but they differ from the normal gabbros in their less coarse texture, in the absence of diallagic and other 'schiller' structures, and in the mutual relations of the felspar and augite which are their two chief constituents. In these respects there is, however, a transition between the two types of rocks.

The diabases occur as large dykes, sills, and laccolitic or other masses. Minor intrusions of rocks having a similar chemical composition commonly have more of the petrographical characters of volcanic rocks. For these we shall retain the names dolerite, andesite, basalt, *etc.*, and they will be excluded from this place.

The name diabase has been, and still is, employed in different senses. By the German school it is usually restricted to the older rocks, whether intrusive or volcanic, dolerite and basalt being terms reserved for rocks of Tertiary or later age. Allport shewed very conclusively that such a distinction corresponds to no real difference between the older and the newer rocks, and he abandoned the name diabase in favour of dolerite for all. The rocks so designated by Allport include some of the intrusive and others of the volcanic types. English writers have followed him in admitting no criterion of geological age into their classification and nomenclature, but some of them have inconveniently employed the name diabase for a more or less decomposed dolerite.

According to the absence or presence of the basic silicate olivine, the rocks of the present family are often divided into *diabases proper* and *olivine-diabases.* Olivine is in general found in the more basic members of the family, but this division does not correspond with any exactness to the chemical division into intermediate (or sub-basic) and basic. By the presence of some other special mineral we may distinguish such types as *quartz-diabase*, *bronzite-diabase*, and *hornblende-diabase*; or again quartz-bronzite-diabase and olivine-hornblende-diabase.

Various other names have been used for particular types of diabasic rocks. Among the hornblende-bearing diabases of the Fichtelgebirge von Gümbel distinguished two types; proterobase, containing original hornblende in addition to augite, and epidiorite, in which the hornblende is all derived from augite. Some writers have extended these names to cover all diabasic rocks characterized by primary and secondary hornblende respectively. The old field-term 'greenstone,' referring to the staining of the rocks by chloritic and other decomposition-products, included not only diabases but diorites, picrites, altered dolerites, *etc.*, and so had no precise signification.

The picrites, included above among the plutonic rocks, have much in common with the diabases, and in some districts are closely associated with them.

Constituent Minerals. The *felspars* of the diabases range from oligoclase to anorthite in different examples: varieties of labradorite are perhaps the most common. The crystals have a strong tendency to idiomorphism, with columnar or sometimes tabular habit. Twin-lamellation on the albite law is universal, and is often combined with carlsbad twinning, but the pericline law is not so common. Zonary growth is not often shewn, except when a later set of felspars occurs, of shapeless outline and more acid composition; these shew strong zoning between crossed nicols. Inclusions are not common, except glass-cavities and needles of apatite. Decomposition gives rise to calcite-dust, to finely divided material, which may be mica, to zeolites, or to granular epidote. The crystals also become charged with strings and patches of green chloritoid substance, probably derived in part from the pyroxene.

The common pyroxenic constituent is an *augite*, usually without crystal outlines. It varies in thin slices from brown to nearly colourless, and rarely shews sensible pleochroism. Zonary and 'hour-glass' structures are sometimes seen. The orthopinacoidal twin is common, and in some cases there is a fine basal lamination[1] in addition (Whin Sill). The commonest decomposition-products are pale green fibrous or scaly aggregates of serpentinous and chloritic substances. The former may be recognized by their low refractive index and moderately high birefringence; the latter are usually very feebly birefringent or sensibly isotropic, and shew distinct pleochroism. Delessite is probably a common product, besides chlorite proper, but the discrimination of this very ill-defined group of minerals is not easy. Another change to which augite is subject is that which results in a light-green 'uralitic' hornblende. This is usually, but not always, fibrous in structure.

Some diabases contain *bronzite* in addition to augite. It is in more or less idiomorphic crystals, with faint pleochroism, and gives rise by alteration to pseudomorphs of light green fibrous bastite.

Only occasionally does *hornblende* appear as an original constituent. It seems to be characteristically a brown variety. Brown *biotite* is also a rare accessory.

A little *quartz* is found in some of the less basic diabases, occurring interstitially. Whether it is original or a decomposition-product is sometimes difficult to decide, but when the

[1] Rosenbusch-Iddings, Pl. XIX, fig. 6; Teall, *Q. J. G. S.* (1884) xl, Pl. XXIX, fig. 1.

mineral forms part of a micrographic intergrowth with felspar its primary nature may safely be assumed.

The *olivine*, which occurs in very many diabases, builds more or less rounded idiomorphic crystals or grains, sensibly colourless or very pale. It has the same mode of alteration as in the olivine-gabbros and peridotites.

The iron-ores, which, in contrast with many gabbros, the diabases contain abundantly, include *ilmenite* and *magnetite*. The two are very commonly associated, and some so-called titaniferous magnetite has been supposed to be a minute intergrowth of the two. They are easily distinguished when they occur as crystals or skeleton-crystals. In most cases the ilmenite has given rise to more or less of its characteristic

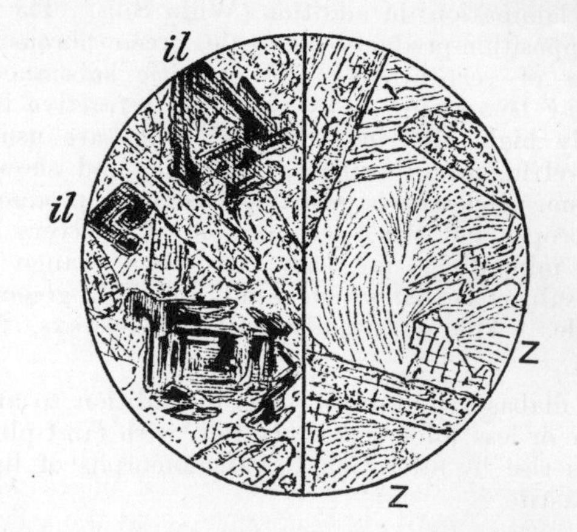

FIG. 24. DECOMPOSING DIABASE, DENEIO, NEAR PWLLHELI, CAERNARVONSHIRE ; × 20.

This shews decomposing felspar-crystals and ophitic augite, with ilmenite-skeletons (*il*), crusted with leucoxene, and patches of radiating fibres of a zeolitic mineral (*z*) [123].

decomposition-product, grey cloudy masses of semiopaque leucoxene[1] (fig. 24).

[1] Rosenbusch-Iddings, Pl. XVI, fig. 2; Teall, Pl. XVII, fig. 2.

Long columnar or needle-like crystals of *apatite* occur in most diabases, but in some are capriciously distributed.

Structure. As regards structure, the diabases offer a contrast to normal plutonic rocks, owing mainly to the fact that the crystallization of the felspar has preceded that of the dominant ferro-magnesian constituent. As seen in a slice, the columnar crystals of felspar shew more or less elongated sections, with no law of arrangement, and around or between these the augite is moulded. The last-named mineral in most cases distinctly wraps round the felspar crystals, and often forms plates of some extent, enclosing many of them. This is known as the *ophitic* structure (fig. 25). In other cases the augite tends to form more or less rounded grains imbedded in a plexus of lath-shaped felspars, adjacent grains not being parts of one crystal but shewing different orientations. This is what Prof. Judd[1] has styled the *granulitic* structure: he considers it due to movement towards the end of the process of consolidation. In both types, if olivine is present it is always idiomorphic towards the augite, but may be penetrated by the felspar prisms. The rhombic pyroxene, too, is constantly of earlier crystallization than the augite, and may shew good outlines. The iron-ores are usually idiomorphic, but magnetite may be in part later than the felspar. When, as is sometimes the case, a subordinate felspar, of later consolidation than the dominant kind, is present, it has crystallized with or after the augite, and is always shapeless.

The typical diabases thus present a very uniform structural character, which in its best development is almost peculiar to them. In a few diabases, however, the augite, especially if not abundant, is partially idiomorphic, and the same is true of rocks which are on the border-line between diabase and gabbro. A porphyritic character, due to the development of relatively large crystals of felspar at an early stage, is not common: it is sometimes connected with an increasing fineness of texture of the rock on approaching the edge of an intrusive mass. Other occasional marginal peculiarities are flow-phenomena, vesicles or amygdules, and the development of

[1] *Q. J. G. S.* (1886) xlii, pp. 68, 76, and figs. Pl. v.

a glassy base or sometimes of variolitic and allied structures. Rocks having these features and occurring as marginal modifications of normal diabases do not differ in any essential from certain types of lavas, and will therefore not be noticed in this place.

Some British Examples. We scarcely need to go beyond our own country to illustrate all the leading types of rocks in this family.

A true *quartz-diabase* is not often met with. In any but quite fresh rocks, at least, it is not possible to be certain that quartz occurring interstitially is really an original constituent of igneous origin. Among the numerous dykes traversing the old gneiss of Sutherland are diabases of which some are quartz-bearing (Loch Glencoul, *etc.*). The chief constituent minerals are a basic plagioclase and a pale or colourless augite, the relations between the two being rather variable. A green or yellow-green hornblende occurs as a marginal alteration of the augite, especially around the grains of magnetite, and a little brown biotite is also associated with the latter. Apatite is the earliest and quartz the latest product of consolidation. The hornblende is connected with mechanical stress in the rock, and specimens may be collected to shew the complete amphibolization of the augite, as well as recrystallization of the felspar. Another feature of these dykes is their fine-textured selvage, which seems in some cases to have been actually glassy.

A well-known rock in the north of England is the Great Whin Sill[1], which is intrusive in Lower Carboniferous strata, and extends from the Northumberland coast to the Eden valley. In its coarser central parts it sometimes approaches a gabbro in aspect, the augite becoming idiomorphic; the fine-textured portions near the margin, on the other hand, take on an andesitic character, developing perhaps some glassy base; but the bulk of the intrusion is of diabase of a distinctive type. The normal structure is more or less ophitic, and the dominant constituents are a lath-shaped felspar, near andesine,

[1] Teall, *Q. J. G. S.* (1884) xl, 640–657, Pl. XXIX: also *Brit. Petr.* Pl. XIII, fig. 2.

and a pale brown augite, often with basal striation. The iron-ore is titaniferous, and may perhaps represent minute intergrowths of magnetite and ilmenite. Apatite occurs sparingly. An accessory mineral is bronzite, tending to be replaced in the usual fashion; brown mica is occasionally seen, and a little brown hornblende is often present, bordering the augite with crystallographic relation. Quartz is detected in all the coarser varieties of the rock, and is at least in part original, since it frequently occurs in micrographic intergrowth with felspar. The rock is thus a *quartz-diabase.* Mr Teall[1] has described a similar rock from Ratho, near Edinburgh.

The Penmaenmawr[2] intrusion, probably of Ordovician age, is also characterized by quartz occurring interstitially in a micrographic intergrowth. In this rock bronzite becomes an essential constituent, being quite as abundant as the pale brown augite. The latter mineral often shews the delicate basal striation already noticed. Biotite is sometimes rather abundant, but the dominant type of rock is a *quartz-bronzite-diabase.* The structure of the rock is rather granulitic than ophitic, and it usually shews some approach to the characters of volcanic rocks in the occurrence of more than one generation of felspar. Some of the latest shapeless crystals are to be referred to orthoclase. The rock passes into a type which would be properly described as an andesite. The general body of the rock is traversed by comparatively coarse segregation-veins of more acid composition[3].

The numerous sills of Ordovician age in Caernarvonshire[4] are of *diabase without olivine*, and have almost universally the ophitic structure. The felspar gives lath-shaped or rectangular sections from ·05 to ·5 inch long, with albite- but only occasionally pericline-lamellation: it often gives extinction-angles indicating labradorite and neighbouring varieties. The augite is pale brown to almost colourless, and very rarely shews any approach to idiomorphism. Besides the commoner decomposition-products, there is often a fibrous colourless hornblende,

[1] Teall, p. 190.
[2] *Bala Volc. Ser. Caern.* 65; Teall, Pl. xxxv, fig. 2.
[3] Waller, *Midland Naturalist* (1885) viii, 1–7.
[4] *Bala Volc. Ser. Caern.* 75–86.

fringing the augite but occupying the place of destroyed felspar, *etc.* The iron-ores include both magnetite and ilmenite, often together, and apatite is locally plentiful. Rhombic pyroxene is wanting, as well as olivine, while original hornblende and quartz are practically absent, and biotite very exceptional. These Caernarvonshire diabases are thus of very simple mineralogical constitution. Despite the absence of olivine, they are of thoroughly basic composition. The diabases of similar age in Wicklow are also free from olivine, and are probably of more acid composition, some of them containing quartz. They are characterized by a partial or even total conversion of the ophitic augite into hornblende, with other changes ascribed to dynamic metamorphism[1].

In the Lake District, diabases are not very largely developed. The rock of Castle Head, Keswick, shews a divergence from the normal type in the presence of porphyritic idiomorphic crystals and crystal-groups of twinned augite. The general mass of the rock has had an ophitic structure, but is much decomposed, with the production of quartz, calcite, a feebly polarising chloritoid substance, and little veins of fibrous serpentine (chrysotile).

Numerous dykes of post-Carboniferous but pre-Permian age are found on the shores of the Menai Straits and in some other parts of Wales and England[2]. The smaller ones are augite-andesites, not calling for any special notice; the larger may be classed as dolerites or as diabases shewing a tendency to a volcanic type. The dominant felspars have the usual rectangular section, and the light brown augite moulds round them in ophitic fashion, but the special feature of the rocks is the occurrence of a second and subordinate generation of felspar in allotriomorphic crystal-grains which have consolidated, on the whole, about simultaneously with the augite. They have less close twin-lamellation than the dominant felspars, are of more acid composition, and always shew a marked zonary banding between crossed nicols. These rocks contain magnetite, but not ilmenite.

[1] Hatch, *G. M.* 1889, 263–265.
[2] *Bala Volc. Ser. Caern.* 109.

Numerous *olivine-diabases* are associated with the Carboniferous strata of the Midlands. Good examples are seen in the Clee Hills, Shropshire[1]. The rock of Pouk Hill, near Walsall, is an ophitic olivine-diabase. In that of Rowley, near Birmingham, the augite occurs in little grains and tends to be idiomorphic[2], or again there is a micrographic intergrowth of augite and felspar[3]. In this rock are relatively acid segregation-veins, in which part of the felspar is orthoclase[4]. A few of the Derbyshire 'toad-stones' have the structure of

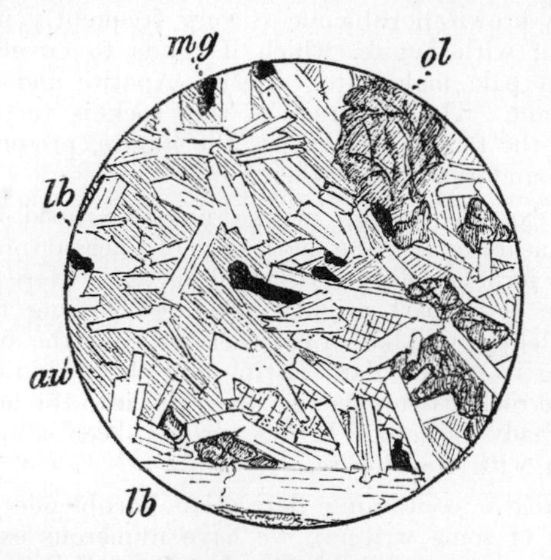

FIG. 25. OLIVINE-DIABASE, BONSALL, DERBYSHIRE; ×20.
Shewing olivine-grains (*ol*), more or less completely serpentinized, magnetite (*mg*), and lath-shaped crystals of labradorite (*lb*), set in a framework of crystalline augite (*au*), which wraps round and encloses the felspar with typical ophitic structure [424].

ophitic diabases (fig. 25), but according to Mr Arnold-Bemrose, they are contemporaneous lavas. (See below, Chap. XIV.)

[1] This and many other British examples were noticed by Allport, *Q. J. G. S.* (1874) xxx, 529–567.
[2] Teall, Pl. XI.
[3] Teall, Pl. XXIII, fig. 2.
[4] Waller, *Midland Naturalist* (1885) viii, 261–266.

The olivine-diabases which occur in the southern part of Scotland and in the islands to the west are in part of Carboniferous, in part of Tertiary age. As distinguished from the basalts and dolerites, they are typically ophitic rocks consisting of magnetite, olivine (often in fresh crystals), lath-shaped felspar, and crystal-plates of augite. Zeolites are frequent among the secondary products.

Of *hornblende-bearing-diabases* a good example is found in a large dyke which runs on the east side of Holyhead Mountain[1]. The brown hornblende is very frequently in parallel intergrowth with augite, which it tends to envelope. The augite is a pale malacolite variety. Apatite and magnetite are abundant. The structure of this rock is very variable, sometimes the felspar, sometimes the augite, presenting idiomorphic boundaries to the other.

Other examples occur in the neighbourhood of Penarfynnydd[2], near Sarn, in the south-west of Caernarvonshire, and apparently form laccolitic masses of Bala age. Here the brown hornblende is in part original, often enveloping the augite with parallel growth, but in part derived from the augite. By the coming in of abundant olivine and the dwindling of the felspar, the rock passes, though abruptly, into the hornblende-picrite already noticed. It has already been alluded to in connection with the diorites.

Of diabases containing derivative hornblende only (the epidiorites of some writers), we have numerous examples in this country. A good one is found at Cuns Fell in the Cross Fell district[3]. This rock has consisted originally of idiomorphic felspar (andesine-labradorite) and colourless augite in ophitic or semi-ophitic plates, with some apatite and magnetite, but it has suffered various secondary changes. These are, in order, (i) the partial or total replacement of augite by greenish yellow uralitic or fibrous, and finally by yellowish brown compact, hornblende, with the usual crystallographic relation; (ii) the growth of colourless hornblende-fringes about both

[1] *G. M.* 1888, 270, 271.
[2] *Q. J. G. S.* (1888) xliv, 450–454; *Bala Volc. Ser. Caern.* 92–97.
[3] *Q. J. G. S.* (1891) xlvii, 524.

augite and hornblende, this proceeding concurrently with alteration of the felspar, *etc.*; (iii) the conversion of much of the remaining augite into chloritoid and other decomposition-products by ordinary weathering action.

Many of the 'greenstones' of Cornwall are much altered diabases shewing uralitization, chloritization, serpentinization, and other changes; but the rocks so named in the field include also old basic lavas and other types[1].

We may briefly notice in this place the peculiar group of rocks, named *teschenite* by Hohenegger, occurring as intrusions in the Cretaceous of Silesia and Moravia (Teschen, Neutitschein, Söhla, *etc.*). They consist mainly of augite, brown hornblende, plagioclase, apatite, and analcime. The augite is often of a violet tint and strongly pleochroic, and it is frequently bordered by hornblende in parallel position. The apatite is very abundant and builds large prisms. The analcime is doubtless secondary, and has been supposed to be derived from nepheline, while some observers have recorded the presence of nepheline in the rocks. Teschenites occur in the Caucasus, in Portugal, *etc.*, and a similar rock is found at Car Craig in the Firth of Forth[2]. It is rich in reddish brown, pleochroic augite, and contains altered felspar, analcime and other zeolites, iron ores, and brown mica (probably secondary). It presents points in common with the neighbouring picrite of Inchcolm. All these rocks are typically non-ophitic, but others more resembling normal diabases have also been included under the name teschenite.

In general, we may use the term teschenite for a nepheline-bearing diabase or for a diabase which, from an abundance of secondary minerals rich in soda, may be supposed to have contained nepheline. Silurian intrusions of this type occur in association with the nepheline-syenite of Montreal. They contain both nepheline and sodalite, and some have olivine.

[1] J. A. Phillips, *Q. J. G. S.* (1876) xxxii, 155–178; (1878) xxxiv, 471–496, Pl. xx–xxii.

[2] Teall, Pl. xxii, fig. 1.

CHAPTER X.

LAMPROPHYRES.

THE lamprophyres are a peculiar group of rocks occurring typically as dykes or other small intrusions. Chemically they are characterized by containing, with a medium or low silica-percentage, a considerable relative quantity of alkalies (especially potash), while the oxides of the diatomic elements are also abundantly represented. This shews itself in the common types of lamprophyres by an abundance of brown mica, and indeed the lamprophyres as a family are rich in ferro-magnesian silicates. They are fine-grained rocks, but almost always holocrystalline, and their structure is in some respects peculiar.

Gümbel's name lamprophyre has been extended by Rosenbusch to cover the various members of this family. The commoner varieties are mica-lamprophyres ('mica-traps', Ger. Glimmertrapp). Of these, two types have long been recognized, a chief point of distinction being the predominance of orthoclase in one and plagioclase in the other. To these types are given the names, respectively, *minette* (a word taken from the miners of the Vosges) and *kersantite* (from Kersanton, near Brest). To these Rosenbusch has added two other types for rocks in which the place of biotite is taken by augite or hornblende. He separates those with dominant orthoclase (*vogesite*) from those with dominant plagioclase (*camptonite*). It should be noted that the criterion of the felspars does not lead in this family to a very natural division, especially when much of the potash in the rocks is present in mica. Further, the decomposition of the rocks often renders the identification of the felspars difficult. For most purposes it is perhaps sufficient to distinguish the rocks merely as mica-, hornblende-, and augite-lamprophyres. There are other types of very restricted occurrence, which will not be specially noticed here. The association with camptonite of the peculiar rock bostonite has already been remarked.

The rocks of this family have a wide range of chemical composition. Their equivalents, from this point of view, among the volcanic types are chiefly basaltic rocks, and especially leucite- and nepheline-bearing basalts. From these the lamprophyres as a whole differ considerably in mineralogical composition, olivine being wanting or poorly represented in many of the types, and the felspathoid minerals occurring only very exceptionally; while, on the other hand, brown mica, a mineral by no means characteristic of basaltic lavas, is a prominent constituent in most of the lamprophyres.

Constituent minerals. The characteristic mineral of those lamprophyres most usually met with is *biotite*, which occurs in hexagonal flakes. The extinction-angle (3° or 4°) is sufficient to shew frequently a lamellar twinning parallel to the basal cleavage. The flakes are very commonly bleached in the interior, retaining only at the margin the normal deep brown colour (fig. 26 *A*). With the bleaching there is a certain diminution in birefringence. More rarely we find a dark interior with a pale border, or a dark nucleus and border with a pale intermediate zone. Complete decomposition results in a pale, feebly polarizing substance as a pseudomorph. A greenish chloritic alteration is also found. Iron-oxide separates out, usually as limonite, and a carbonate (calcite or dolomite) is produced as little wedges or lenticles along the cleavages of the mica (fig. 26 *A*). The titanic acid of the mica separates out as rutile, in fine needles arranged in three sets at angles of 60° : this is well seen in basal sections (fig. 26, *B*). The original inclusions of the biotite are apatite and sometimes magnetite and zircon.

Short columnar crystals of *augite* occur in many lamprophyres, shewing sharp outlines with an octagonal cross-section and sometimes lamellar twinning. When fresh, the mineral is almost colourless in slices, but it is readily replaced by serpentine, calcite, chlorite, *etc.*, in good pseudomorphs. In other cases uralization may be noticed. The augite crystals

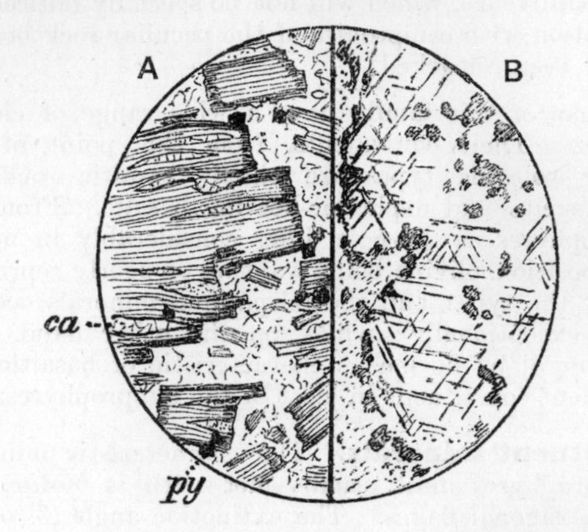

FIG. 26. MICA-LAMPROPHYRE (MINETTE).

A. Helm Gill, near Dent, Yorkshire; ×20: shewing the internal bleaching of the mica-flakes and the formation in them of lenticles of calcite (*ca*). The opaque iron-ore (*py*) is pyrites [444]. *B.* Decomposing biotite in mica-lamprophyre, Budlake, near Exeter; ×100: shewing the production of rutile-needles and patches of limonite [1346].

are sometimes coated with flakes of biotite. The most usual occurrence of *hornblende* is in long well-shaped prisms, frequently twinned, but it has some variety of habit. The colour is brown or sometimes green. The mineral may be converted into a chloritic substance with separation of iron-oxides.

A striking feature in the lamprophyres is that the felspars do not usually occur as phenocrysts. The nature of the felspar in the more altered rocks can be verified only after removing the carbonates from the slice with dilute acid. The small columnar or tabular crystals of *plagioclase* shew albite-lamellation and frequently zonary banding. They often have a kind of sheaf-like grouping. Decomposition, beginning in the interior, gives rise to abundant calcite. The *orthoclase*, and perhaps anorthoclase, build short rectangular crystals, simple or carlsbad twins, often clouded or with ferruginous staining.

Some of the more acid lamprophyres have a certain amount of *quartz*, which is either the latest product of consolidation or is intergrown with a portion of the felspar with micrographic structure.

A common accessory in some lamprophyres, and an essential in certain types, is *olivine*, which builds relatively large perfect crystals, or sometimes groups of rounded grains. It is occa-

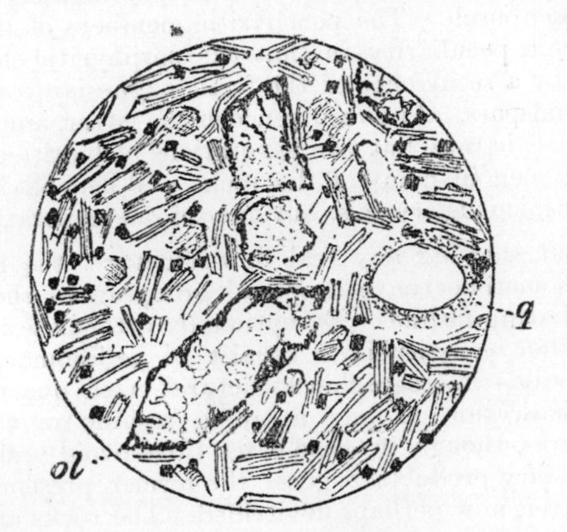

FIG. 27. MICA-LAMPROPHYRE, RAWTHEY BRIDGE, NEAR SEDBERGH, YORKSHIRE; ×20.

There are abundant flakes of biotite, with internal bleaching, and octahedra of magnetite. The part shewn clear is an aggregate of felspar crystals obscured by secondary calcite dust. Olivine is represented by pseudomorphs (*ol*) of calcite coated with red iron-oxide. There is an enclosed grain of quartz (*q*) with a corrosion-border of augite, now decomposed [1598].

sionally found fresh, but very commonly represented by pseudomorphs of carbonates and serpentine (fig. 27). Olivine is not found in the ground-mass.

The iron-ores are not often very abundant, and may be

quite wanting. The most usual is *pyrites*, but octahedra of *magnetite* are also found.

A constant and abundant accessory is *apatite*, but it is sometimes in such fine needles as to be invisible except by oblique illumination. Sphene and zircon are only exceptionally met with.

Structures and peculiarities. Many of the lamprophyres are non-porphyritic, with a rather exceptional structure due to a strong tendency to idiomorphism of all the constituent minerals. This is the panidiomorphic structure of Rosenbusch. The porphyritic members of the family, again, have a peculiarity, in that the porphyritic character is produced by a recurrence of the ferro-magnesian constituents, not of the felspars. Any recurrence of the latter, and especially of orthoclase, is rare, but two generations of biotite or of hornblende are seen in many of the rocks. When olivine occurs, it is in conspicuous crystals, but only of one generation.

Without shewing any real flow-structure, the felspars of the rock sometimes have a special grouping in sheaf-like or rudely radiating fashion. Exceptionally orthoclase is moulded on the other constituents: usually it is idiomorphic, save when it builds micrographic structures with quartz. There is little indication of any isotropic residue in the typical lamprophyres, though in some cases little ovoid vesicles, filled with secondary products, suggest the former presence of some glassy matter, now perhaps devitrified. The rocks are remarkably prone to decomposition, and often have 20 or 30 per cent. of calcite and other secondary products.

Grains of quartz and crystals of alkali-felspars are found, though very sparingly, in many lamprophyres. Their sporadic occurrence and, still more, some curious features which they invariably present compel us to regard them as something apart from the normal constitution of the rock and of quasi-foreign origin. The *enclosed quartz grains* (fig. 27) are of rounded form with evident signs of corrosion, and are seen to be surrounded by a narrow ring or shell due to a reaction between the quartz and the surrounding magma. This shell is probably in the first place of augite, but it is often found to consist of minute flakes of greenish fibrous hornblende or of calcite and chloritoid products. The quartz having this mode of occurrence must be distinguished from genuine derived fragments torn from other rocks: these are of irregular form, often complex, and may contain inclusions unknown in the corroded quartz-grains. The two may sometimes be seen in the same slide.

The *enclosed felspar crystals* are always of an acid species —either orthoclase or a plagioclase rich in soda. The crystals are corroded so as to present a rounded outline, but not re-

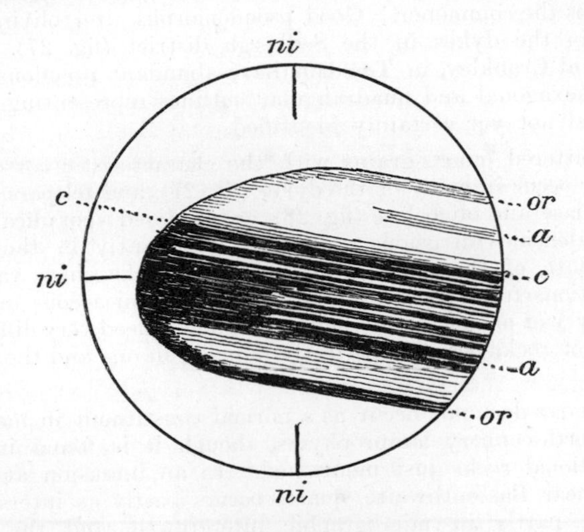

FIG. 28. OLIGOCLASE CRYSTAL ENCLOSED IN A LAMPROPHYRE DYKE AT GILL FARM, NEAR SHAP WELLS; ×20.

Crossed nicols. The crystal is rounded by magmatic corrosion and bordered by a narrow margin of orthoclase (*or*). In addition to the albite-lamellation of the oligoclase (*a*), there is a carlsbad twinning (*c*) common to both felspars [1155].

duced to mere round grains. The plagioclase thus corroded is bordered by a narrow margin of orthoclase due to the action of the magma (fig. 28).

Illustrative examples. The best-known British examples occur as small dykes and sills in the north of England[1], and are of an age between the Silurian and the Carboniferous. The dykes are numerous in the southern part of the Lake District from Windermere to Shap and on to Sedbergh, and they are seen again in the Lower Palæozoic inliers of Ingleton, Edenside, and Teesdale. The rocks are mica-lamprophyres, but many of them contain subordinate augite, always in perfect crystals, but often decomposed. The relative proportions of orthoclase and plagioclase vary, so that some examples would be named minette and others kersantite, the latter being perhaps the commoner. Good pseudomorphs after olivine are seen in the dykes in the Sedbergh district (fig. 27). The dykes at Cronkley, in Teesdale, have abundant pseudomorphs with hexagonal and quadrangular outlines representing some mineral not yet certainly identified.

Scattered quartz-grains with the characteristic corrosion-border occur in many of the dykes (fig. 27), and felspars, both orthoclase and oligoclase (fig. 28), are enclosed sporadically in the Edenside intrusions and more abundantly in those to the south of the Shap granite. These rocks shew various transitions from typical lamprophyres to a micaceous quartz-porphyry of one of the less acid types, and indeed very different kinds of rocks occur imperfectly mingled in one and the same dyke.

Quartz does not occur as a normal constituent in most of the north-country lamprophyres, though it is found in the transitional rocks just mentioned. In an intrusion at Sale Fell, near Bassenthwaite, quartz occurs partly as interstitial grains, partly in micrographic intergrowth, and the rock shews considerable resemblance to the original kersantites of Brittany[2]. The last-named rocks are sometimes even-grained, sometimes porphyritic ('porphyrites micacées' of Barrois).

An augite-bearing minette seems to be one of the commonest types of lamprophyres. It is seen in Cornwall (Trelissick Creek, *etc.*), in the Channel Islands (Doyle Monument, Guernsey),

[1] *G. M.* 1892, 199–206, with numerous references.
[2] Fouqué and Lévy, Pl. IX.

and at numerous foreign localities (*e.g.* Plauen'scher Grund, near Dresden). With more abundant augite (*e.g.* Weinheim in the Odenwald) it passes into the augite-vogesites. The vogesites of the Vosges, *etc.*, have sometimes augite, sometimes hornblende as the dominant coloured constituent.

The numerous lamprophyres of Scotland are for the most part not yet described. Some of the rocks occurring as sills in the Assynt district of Sutherland[1] seem to be rather camptonites than diorites. They are characterized by green horn-

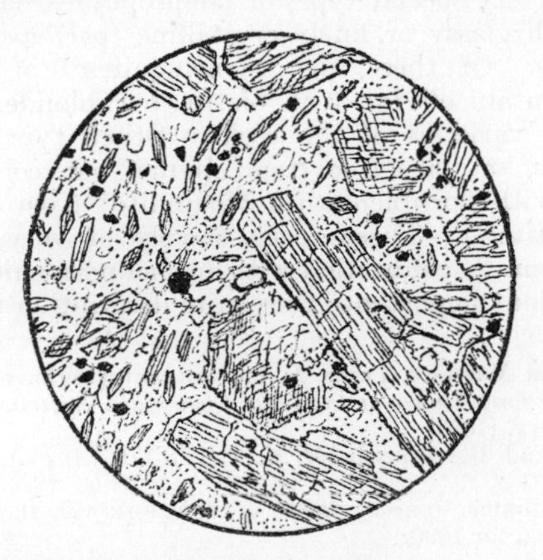

FIG. 29. HORNBLENDE-LAMPROPHYRE (APPROACHING CAMPTONITE), FROM INTRUSIVE SILL IN DURNESS LIMESTONE, LOCH ASSYNT; ×20.

Shews phenocrysts of green hornblende in a panidiomorphic ground-mass of plagioclase and hornblende, with a little magnetite and apatite [1687].

blende in rather slender twinned prisms (fig. 29). One sill, however, is a mica-lamprophyre with accessory augite and olivine (both destroyed).

The more typical camptonites, such as those of the north-eastern United States, are characterized by brown hornblende, which makes up the greater part of the rock, with subordinate

[1] Teall, *G. M.* 1886, 346–353.

augite and oligoclase. In several districts such rocks occur as dykes in evident relation with masses of nepheline-syenite, and it may be noticed that the sills of the Assynt district are associated with the singular rock of Loch Borolan. The mica-lamprophyres, on the other hand, seem to be related to granites. They are often associated with acid dyke-rocks, such as micro-granite or even aplite. Similarly the camptonite dykes are frequently accompanied by dykes of a peculiar rock of trachytic nature (bostonite), consisting almost wholly of alkali-felspars[1].

Certain very special types of lamprophyres have been described, with glassy or finely crystalline (perhaps devitrified) ground-mass. Of these, the 'monchiquites[2]' of Brazil and Portugal contain olivine with biotite, hornblende, and augite in different varieties of the rocks. Other types, occurring in Arkansas, *etc.*, are free from olivine, but contain augite associated with hornblende (in 'fourchite[3]') or with biotite (in 'ouachitite[4]'). These rocks, mostly of very basic composition, occur in connection with nepheline-syenites, and are not of sufficient importance to require description here.

[1] Kemp and Marsters, *Trans. N. Y. Acad. Sci.* (1891), vol. xi, pp. 13–23. Also *The Trap Dykes of the Lake Champlain Region*, Bull. No. 107 *U. S. Geol. Surv.* (1893).

[2] Hunter and Rosenbusch, *M. M.* x, 177–178 (*Abstr.*). Also Kemp and Marsters, *l.c.*

[3] J. F. Williams, *Igneous Rocks of Arkansas*, vol. ii, of *Ann. Rep. Geol. Surv. Ark.* for 1890.

[4] Kemp *ibid.*

C. VOLCANIC ROCKS.

UNDER this head we shall treat only the solid rocks of volcanic origin (lavas), reserving the fragmental products of volcanic action for the sedimentary group. With the true extruded lava-flows will be included similar rocks occurring in the form of dykes, *etc.*, in direct connection with volcanic centres, the common feature of all being that they have consolidated from fusion under superficial conditions, *i.e.* by comparatively rapid cooling under low pressure. This mode of origin has given the rocks as a whole characters which place them in contrast with the plutonic group, while the types treated above under the head of 'intrusive' have in some respects intermediate characters. Many volcanic outpourings have undoubtedly been sub-marine, and when these have taken place under a great depth of water the products may be expected to approximate in some measure to the characters of rocks of deep-seated origin. On this point we have not much certain information, but quartz-porphyry and diabase lava-flows are known, and certain extensive sheets of gabbro in the Lake Superior region have been stated to be volcanic outflows. In general, however, the contrast between volcanic and plutonic types of structure is well marked.

The presence of a glassy (or devitrified) residue, though not peculiar to volcanic rocks, is highly characteristic of them, and especially of the more acid members. Other features characteristic of lavas, though not confined to them, are the vesicular and amygdaloidal structures, and the various fluxion-phenomena, including flow-lines, parallel orientation of phenocrysts, banding, drawing out of vesicles, *etc.*

The great majority of the volcanic rocks have a porphyritic structure, *i.e.* their constituents belong to two distinct periods of consolidation, the earlier represented by the porphyritic crystals or 'phenocrysts'[1], and the later by the 'ground-mass,' which encloses them, and commonly makes up the bulk of the rock. This ground-mass may, and usually does, include some glassy residue or 'base': if the ground is wholly glassy, we have what is termed the 'vitrophyric' structure. The same mineral—say quartz or augite—may occur both among the phenocrysts and as a constituent of the ground-mass. When such a recurrence is found, the crystals of the earlier generation are distinguished from those of the later by their larger size, often by their more perfect idiomorphism, and in some cases by fracture, corrosion, or other evidence of vicissitudes in their history. The two periods of consolidation are styled by Rosenbusch the 'intratelluric' and the 'effusive,' the former being considered as the result of crystallization prior to the pouring out of the lava, and so under more or less deep-seated conditions. When we speak of the consolidation of a lava at the earth's surface, we must be understood to refer to the ground-mass of the rock. In some few types of lavas the phenocrysts fail altogether, and the effusive period is the only one represented.

The various types will be grouped under families, to be taken roughly in order beginning with the most acid. It is customary to speak of the several families of lavas as answering to the commonly recognized families of the plutonic rocks —the rhyolites to the granites, the trachytes to the syenites, *etc.*—but such a correspondence cannot be followed out with great exactness. It is certain that a given rock-magma may result in very different mineral-aggregates according as its consolidation is effected under deep-seated or under surface conditions; and in the latter case, moreover, much of the rock produced may consist of unindividualised glass.

It is more especially in the volcanic rocks that the Continental petrologists have insisted upon a division into an

[1] This convenient term, due to Prof. Iddings, will be adopted here. Mr Blake has proposed the word 'inset,' as corresponding to the Ger. 'Einsprengling.'

'older' and a 'younger' series ('palæovolcanic' and 'neovolcanic'), an arbitrary line being drawn between the pre-Tertiary lavas and the Tertiary and Recent. This distinction is rejected by the British school, and will find no place in the following pages[1]. The simplified grouping of the volcanic rocks by their essential characters, without reference to their age or supposed age, involves some modification of the double nomenclature in use among the German and French writers. The names employed by them for the younger lavas only will here be extended to all rocks of the same character, irrespective of their geological antiquity.

[1] On this question see *Sci. Progr.* (1894), ii, 48–63.

CHAPTER XI.

RHYOLITES.

THE rhyolite family includes all the truly acid lavas; rocks of porphyritic or vitrophyric structure, in which alkali-felspars and frequently quartz figure as the chief constituent minerals. By the older writers most of these rocks were included, with others, under the large division 'trachyte.' The present family was separated by Richthofen with the name 'rhyolite,' expressing the fact that flow-structures are commonly prominent in the rocks. Roth used the term 'liparite' in nearly the same sense. The Continental petrographers, following their regular principle, use these names for the Tertiary and Recent acid lavas only, the older (pre-Tertiary) being more or less arbitrarily separated and designated by other names (quartz-porphyry, porphyry, *etc.*); and English geologists have often tacitly adopted a like division, calling the older rhyolites, which have often suffered various secondary changes, quartz-felsites, felsites, *etc.*

Some geologists distinguish between potash- and soda-rhyolites, according to the predominance of one or the other of the alkalies, but in rocks which are largely glassy this difference does not always express itself in the minerals present. There is, however, a peculiar group of acid lavas very rich in alkalies, and especially in soda: these rocks, the 'pantellarites' of Förstner, contain special characteristic minerals.

We shall consider briefly the characters of the phenocrysts or enclosed crystals and of the ground-mass. In some rhyolites the phenocrysts occur only sparingly, or may even fail altogether.

Phenocrysts. Among the phenocrysts or porphyritically included crystals of the rhyolites, the most constant are alkali-felspars; both *orthoclase* (including *sanidine*) in tabular or columnar crystals, simple or twinned, and an acid plagioclase, ranging from *albite* to *oligoclase*, in tabular crystals with the usual twin-lamellation. A parallel intergrowth of the monoclinic and triclinic species is occasionally found. The felspars often contain glass- and gas-cavities, but rarely fluid-pores: such minerals as apatite, magnetite, biotite, *etc.*, may be sparingly enclosed. Certain rocks specially rich in soda (pantellarites, *etc.*) have *anorthoclase.*

Quartz, when present, occurs in dihexahedral crystals, often corroded and with inlets of the ground-mass. Besides occasional inclusions of minerals of early consolidation, it contains glass- but rarely fluid-cavities.

The more basic minerals are not present in great abundance. The most usual is *biotite* in deep-brown hexagonal flakes, with only occasional inclusions of apatite, zircon, or magnetite. A greenish *augite* with octagonal cross-section may be present, but brown *hornblende* is much less common. The pantellarites have the brown triclinic amphibole *cossyrite*, with intense absorption and pleochroism.

The most usual iron-ore is *magnetite*, but it is rarely abundant. Needles of *apatite* and minute crystals of highly refringent and birefringent *zircon* may also occur in small quantity. In rarer cases *garnet* is found instead of a ferro-magnesian bisilicate.

Ground-mass and structures. The rhyolites exhibit in their ground-mass a great variety of texture and structure. The texture may be wholly or partly glassy; or cryptocrystalline, often with special structures; or, again, evidently crystalline, though on a minute scale. Further, these several varieties of ground-mass may be associated in the same rock and in the same microscopical specimen. Fluxion is frequently marked by banding, successive bands being of different textures, so that thin layers of glassy and stony or spherulitic nature alternate with one another.

The *vitreous* type of ground-mass alone is found in the obsidians[1]. These rocks, colourless or very pale yellow in thin slices, afford good examples of structures common to all

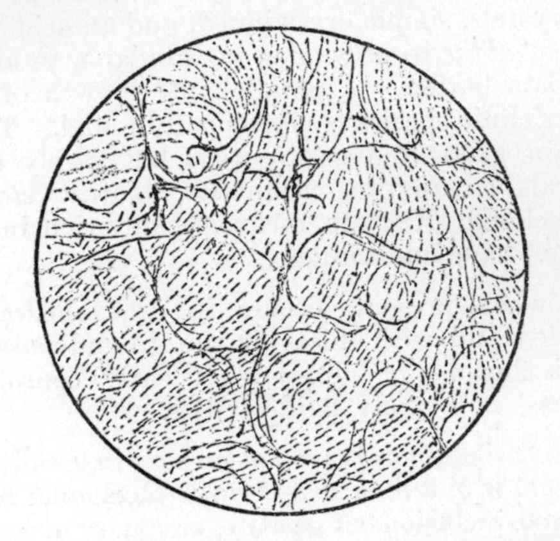

FIG. 30. GLASSY RHYOLITE (OBSIDIAN), TELKIBANYA, NEAR SCHEMNITZ, HUNGARY; ×20.

Shewing sinuous flow-lines traversed by a system of curving perlitic fissures [G. 329].

the natural glasses; especially the *perlitic cracks* (fig. 30), produced by contraction of the homogeneous material[2], and the *vesicular* structure due to the rock-magma having been distended by steam-bubbles. In extreme cases the cavities are so numerous as to make up the chief part of the volume of the rock, and we have the well-known pumice (Fr. ponce, Ger. Bimstein). The vesicles are commonly elongated in the direction of flow, and may even be drawn out into capillary

[1] The less common glassy rocks of the trachyte and phonolite family are also termed obsidian. They are not easily distinguished from the rhyolite-glasses. Some of the rocks styled pitchstones are lavas of the obsidian type, usually of acid composition.

[2] On artificial production of perlitic structure see Cole, *G. M.* 1880, 115–117; Chapman, *ibid.* 1890, 79, 80.

tubes. In the older lavas vesicles are usually filled by secondary products, and become amygdules.

In many cases a ground-mass consisting essentially of glass

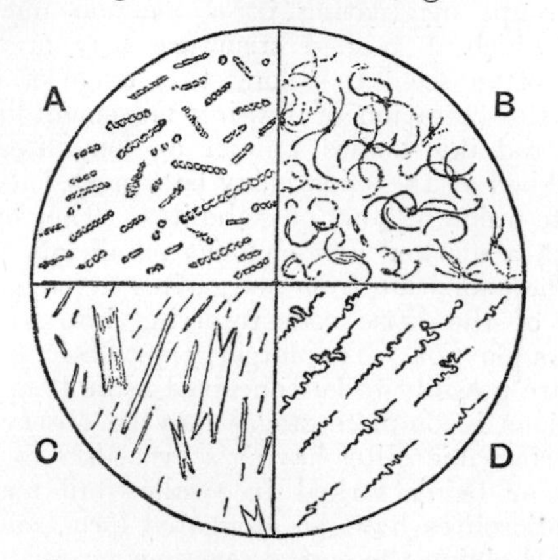

FIG. 31. CRYSTALLITES IN OBSIDIAN.

A. Margarites, Obsidian Cliff, Yellowstone Park; ×400 [477]. *B.* Trichites, Telkibanya, Hungary; ×100 [G. 327]. *C.* Longulites and swallow-tailed crystallites, Hlinik, Hungary; ×200 [G. 70]. *D.* Flow-structure marked by arrangement of twisted trichites, Prabacti, Java; ×200 [G. 64].

still encloses minute bodies known as *crystallites* (fig. 31), which may be regarded as embryonic crystals[1]. They have definite forms, but no perfect crystal boundaries, and the more rudimentary types cannot be subjected to optical tests to determine their nature. The simplest effort at individualisation from the vitreous mass results in globulites, minute spherical bodies without action on polarized light. They occur in profusion in many obsidians, either uniformly distributed or aggregated into cloudy patches (cumulites). From the partial coalescence of a series of globulites arranged in a line result margarites, resembling a string of pearls. A high-power objective (say $\frac{1}{8}$ inch) is often necessary to resolve this beaded

[1] See Rutley, *M. M.* (1891) ix, 261–271, and Plate; Zirkel, *Micro. Petr. Fortieth Parallel*, Pl. IX, figs. 1–4; Rosenbusch-Iddings, Pl. II, III.

structure. Long threads of this nature may extend in the direction of flow but with numerous little twists[1]. Similar threads with curved hair-like form, known as trichites, often occur in groups originating in a common nucleus. These bodies, in which a beaded structure may or may not be observable, often seem to belong to a stage of development later than the cessation of flowing movement in the mass[2]. The small rod-like bodies known as longulites, sometimes slightly clubbed at the ends[3], may be regarded as built up by the complete union of rows of globulites. They often occur in crowds, with a marked arrangement parallel to the direction of flow. The transition from margarites to longulites is often seen, some of the little rods resolving into beaded strings, while others do not. The larger crystallitic bodies termed microlites are possibly to be conceived as built up from longulites. Various incomplete stages may be observed, the ends of the imperfect microlites having a brush-like form (scopulites of Rutley) or being forked in swallow-tail fashion. Fully developed microlites have an elongated form, and are indeed small crystals giving the optical reactions proper to the mineral (felspar, augite, hornblende, *etc.*) of which they consist.

An original *cryptocrystalline* or 'microfelsitic' ground-mass is found in some rhyolites, though it seems to be more characteristic of intrusive types (approaching what we have styled quartz-porphyries) than of true surface lavas. It consists in a granular mixture of felspar and quartz on so minute a scale that the individual grains cannot be resolved in a thin slice. There is no doubt, however, that in many old acid lavas a cryptocrystalline ground-mass has resulted from the *devitrification* (Ger. Entglasung) of a rock originally vitreous. No single criterion can be set up for distinguishing an original from a secondary cryptocrystalline structure. In a rock otherwise fresh, however, there will generally be no reason to suspect devitrification; while, on the other hand, the presence of perlitic cracks is often taken to indicate that the rock in which they occur was originally glassy.

[1] Zirkel, *Fortieth Parallel*, Pl. IX, figs. 3, 4.
[2] *Ibid.* figs. 1, 2.
[3] Fouqué and Lévy, Pl. XVI, fig. 2.

A *microcrystalline* (as distinguished from cryptocrystalline) ground-mass is not very prevalent in true acid lavas, but may occur as bands alternating with glassy or micro-spherulitic bands, often on a small scale. When an evident microcrystalline structure has been set up as a secondary alteration, it probably indicates, as a rule, something more than the merely physical change of devitrification. It is often connected with an introduction of silica from an external source, and in the resulting microcrystalline mosaic quartz often plays a more important part than it would do in a normal igneous rock. In some of the partly silicified Ordovician rhyolites of Westmorland a secondary quartz-mosaic still shews clear indication of former perlitic cracks, outlined by dust, as well as the characteristic banding. In these rocks, too, *silicification* has sometimes affected not only the ground-mass but the felspar phenocrysts.

Another change which has sometimes affected the ground-mass of rhyolites, as well as the felspar phenocrysts, is that which is characterized by an abundant production of epidote[1].

Spherulitic and allied structures. The spherulitic growths which are common in many acid lavas may be conveniently divided into the larger and the smaller. Under the former head we have spherulites, often isolated, with diameters ranging from a fraction of an inch to several inches. They are best studied in certain obsidians, where they are usually of distinctly globular form and with well-defined boundary. The examples which have been most carefully examined, and may be taken as typical, consist mainly of extremely delicate fibres of felspar, arranged radially and on the whole straight, but often forked or branching. In the spherulites of perfectly fresh rocks the space between the fibres is found to be occupied in great part by aggregates of tridymite. In older spherulites, where tridymite is not recognized, quartz may perhaps be considered to represent it. In any case the structure is to be made out only in carefully prepared and very thin slices. It

[1] *E.g.* Rutley, *Q. J. G. S.* (1888) xliv, 740–744, Pl. XVII. Most of the so-called 'epidosites' (epidote-quartz-rocks) are, however, decomposed diabases, *etc.* The name was first used by Pilla for rocks of this nature associated with the gabbros of Tuscany.

may often be observed that the flow-lines of the lava pass undisturbed through the spherulites, indicating that the latter crystallized after the cessation of movement. Spherulites are

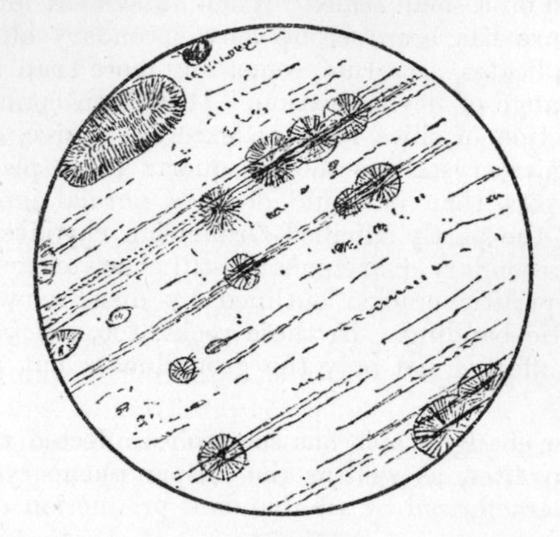

FIG. 32. OBSIDIAN, VULCANO, LIPARI IS.; ×20.
The glassy matrix encloses isolated spherulites, with some tendency to coalesce in bands following the direction of flow. The flow-lines pass uninterruptedly through the spherulites [1785].

often developed along particular lines of flow, and may coalesce into bands (fig. 32).

These larger spherulites shew many special peculiarities in different examples. Sometimes their outward extension has been effected in two or more stages, which are marked by a change in the character of the growth. Again, curious phenomena arise from the formation of shrinkage-cavities (*lithophyses*) in connection with spherulites. Some remarkable examples of lithophyses have been described from the Yellowstone Park[1] and other districts in the United States, from Hungary, and from Lipari[2]. A peculiar feature is the oc-

[1] Iddings, *Obsidian Cliff*, in *7th Ann. Rep. U. S. Geol. Surv.* (1888) 265, 266, Pl. XII–XIV.

[2] Cole and Butler, *Q. J. G. S.* (1892) xlviii, 438–443, Pl. XII; Johnston-Lavis, *G. M.* 1892, 488–491.

currence in the hollows of perfect crystals of the iron-olivine (fayalite), as well as aggregates of tridymite, and in some cases crystals of garnet, topaz, *etc.* The complex forms of these lithophyses can be realised only from specimens or figures. They must be distinguished from ordinary ovoid vesicles.

The large spherulites are in some cases only skeleton-structures, the divergent rays being imbedded in glass. Such *skeleton-spherulites*, in a devitrified matrix, have been described by Prof. Cole[1] in the 'pyromérides' of Wuenheim, in the Vosges.

Examination of the older acid lavas shews that the large spherulites are specially susceptible to certain chemical changes. They are often found partly or totally replaced by flint or quartz, while their insoluble decomposition-products remain as roughly concentric shells of a chloritoid or pinitoid substance. Again, a central hollow is often found, and it is not always clear whether this is due entirely to decomposition or partly represents an original lithophysal cavity.

FIG. 33. MICROSPHERULITIC RHYOLITE, GREAT YARLSIDE, WESTMORLAND; ×20.

Crossed nicols. Each little spherule shews a black cross [1813].

[1] *G. M.* 1887, 299–303.

The very minute spherulites commonly occur in large numbers, closely packed together, so as to constitute the chief bulk of particular bands, or even of the whole ground-mass of the rock. This is the *microspherulitic* structure[1]. The true nature of these very minute bodies, as composed of fine fibres of felspar with quartz, is a matter rather inferred than seen in any given case, but the radiate growth is detected by means of the 'black cross' which each individual spherulite shews between crossed nicols (figs. 33, 34 *A*). These minute spherulites seem to be much less readily destroyed than the larger ones. The *axiolites* of Zirkel[2] seem to be of the nature of elongated spherulites, the fibres radiating not from a point but from an axis (fig. 34 *A*), or they may be conceived as representing the coalescence of a row of minute spherulites (*cf.* fig. 32).

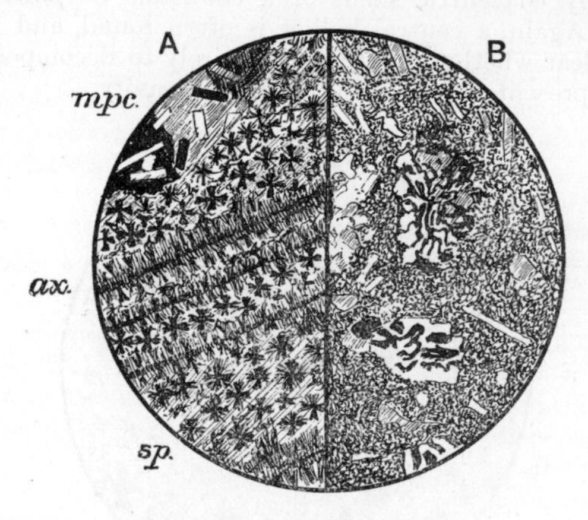

FIG. 34. SPECIAL STRUCTURES IN RHYOLITES, ×20.

Crossed nicols. *A*. Falls of Gibbon River, Yellowstone Park: different bands, following the flow-lines, shew micropœcilitic (*mpc*), axiolitic (*ax*), and microspherulitic (*sp*) structures [1430]. *B*. Goodwick, near Fishguard, Pembrokeshire; shewing micropegmatite phenocrysts in a finely microcrystalline ground-mass.

[1] See Teall, Pl. XXXVIII.

[2] *Micro. Petr. Fortieth Parallel*, Pl. VI, fig. 2. But compare Cole, *M. M.* (1891) ix, 271–274.

Any evident micrographic structure is not common in the ground-mass of rhyolites, though bands or streaks having this character are sometimes found. A curious feature, first described by Prof. Iddings in some obsidians from the Yellowstone Park[1] and rhyolites from the Eureka district of Nevada[2], is the occurrence of porphyritic 'granophyre groups' or *micropegmatite phenocrysts* in a glassy, cryptocrystalline, or microcrystalline ground-mass (see fig. 34 *B*). In these the quartz is subordinate to the felspar in quantity, and the micrographic groups often shew the crystal-boundaries of the latter mineral. As a rule, however, there are several felspar crystals grouped together, the whole permeated by wedges of quartz, and the outline is complex or rather irregular.

A structure met with in the ground of some rhyolites, and in some bands in laminated rhyolites, differs essentially from the micrographic, in that it indicates the successive, instead of simultaneous crystallization of the two constituent minerals. Minute felspar crystals with no orderly arrangement are enclosed in little ovoid or irregular areas of quartz, the whole of the quartz in such a little area being in crystalline continuity. This structure reproduces on a minute scale the ophitic and pœcilitic structures presented by different minerals in other rocks, and Prof. G. H. Williams has adopted for it the term *micropœcilitic*[3] (fig. 34, *A*).

A holocrystalline texture on other than a minute scale is rarely, if ever, met with in true rhyolites. The 'nevadite' of Richthofen comes here, and is exceptional in that the ground-mass is quite subordinate in quantity to the crowded phenocrysts.

Examples of Leading types. The glassy type (obsidian) is exemplified by many of the rhyolites of Iceland and of Lipari (fig. 32), and in the latter locality pumice is

[1] *Obsidian Cliff*, in 7th *Ann. Rep. U.S. Geol. Surv.* (1888) 274–276, Pl. xv.
[2] *Geol. of Eureka District*, Monog. xx, *U. S. Geol. Surv.* (1893) 375, Pl. v, fig. 2.
[3] *Journ. of Geol.* (1893), i, 176–179.

extensively developed. The Hungarian rhyolites are not usually obsidians, but some good examples occur (Telkibanya) with a rich variety of crystallites (fig. 31). Other well-known obsidians come from Ascension Is., Mexico, and the Yellowstone Park. The rock of Obsidian Cliff in the last-named district frequently contains spherulites of some size, isolated or in bands, and remarkable chambered lithophyses, in which occur nests of tridymite and little crystals of the iron-olivine (fayalite). Very similar phenomena have been described from Lipari (Rocche Rosse)[1], and some of the Hungarian lavas also contain small lithophyses, often of hemispherical form, cut off by the fluxion-banding of the lava. It was there that these curious structures were first observed by von Richthofen (Telkibanya, Göncz, *etc.*).

The more widely distributed types of rhyolites may be studied in rich variety from the Tertiary volcanic districts of Schemnitz in Hungary, of the Lipari group, of the Western States of America, *etc.* They differ in the nature of their phenocrysts and in the structure of their ground-mass. Many of them have a strongly marked banded structure, successive narrow bands, a fraction of an inch wide, being of different textures or structures (glassy, microspherulitic, axiolitic, microcrystalline, micropœcilitic). The most usual ferro-magnesian mineral is biotite, but it is never plentiful.

When spherulitic structures are present they may be on a more or less minute scale. Some flows in the Schemnitz district are built up almost wholly of very diminutive spherulites[2], each giving a perfect black cross (Telkibanya, Sarospatak, Eisenbach, *etc.*). In the typical 'perlites' of the same district the individual spherulites are larger, with well-marked radial fibrous structure and globular form, sharply bounded, often by perlitic fissures (Hlinik, *etc.*). These contrast with a type in which the spherulites have an irregular outline, interlocking with one another or sending out processes into a glassy matrix.

[1] Cole and Butler, *Q. J. G. S.* (1892) xlviii, 438–445; Johnston-Lavis, *G. M.* 1892, 488–491.

[2] Fouqué and Michel Lévy, *Min. Micr.*, Pl. XVII, fig. 1.

Of fresh Tertiary rhyolites good examples occur in Antrim, the Tardree rock being well known. It is conspicuously porphyritic, with a soda-bearing sanidine, an acid plagioclase, and corroded quartz-grains. Biotite and magnetite are also present. The ground-mass seems to be here cryptocrystalline to microcrystalline. In this rock von Lasaulx[1] found scales of tridymite, chiefly in little nests occupying cavities in the rock. In specimens from Sandy Braes Mr Watts[2] has remarked perlitic cracks traversing in common the ground-mass (here a brown glass) and the porphyritic crystal-grains of quartz. From the island of Eigg comes, beside other types, a sub-acid obsidian with only scattered phenocrysts of sanidine. The fresh glassy matrix contains few crystallitic growths, chiefly small 'cumulites,' and shews fine perlitic fissuring.

The most interesting British rhyolites, however, are those belonging to the Palæozoic and older volcanic groups, and these have doubtless had their pristine characters modified in many instances by secondary physical and chemical changes.

The Ordovician rhyolites of Caernarvonshire[3] are characterized by the general paucity of any phenocrysts, and especially those of quartz[4]. Among the scattered felspar-crystals, a member of the albite-oligoclase series predominates over orthoclase. Almost the only ferro-magnesian constituent is a little colourless augite, and this is commonly wanting, though a pale green decomposition-product may perhaps represent it. In all these features the rocks closely resemble the Tertiary and Recent rhyolites of Iceland[5], and probably the older rocks have once been largely glassy, as the younger are now. The usual texture of these old lavas is cryptocrystalline to microcrystalline, sometimes shewing fluxion and banding, and occasionally good perlitic cracks. The vesicular structure is not very frequent. In some types the ground is partly micro-

[1] *Journ. Geol. Soc. Irel.* (1878) xiii, 25–31.
[2] *Q. J. G. S.* (1894) l, 367–375, Pl. XVIII.
[3] *Bala Volc. Ser. of Caern.* 18–23.
[4] This is true more especially of central and eastern Caernarvonshire. The rhyolites of the Lleyn peninsula, many of which are intrusive, are richer in phenocrysts, including quartz.
[5] Bäckström, *M. M.* (1894) x, 343, 344 (*Abstr.*).

pœcilitic, minute felspar prisms being enclosed in quartz (Penmænbach, *etc.*). Any approach to a microspherulitic structure of a perfect type is uncommon, but large isolated spherulites are abundant in many localities, and shew the various secondary alterations, concentric shell-structure, silicification, *etc.*, to which they are always prone[1]. The siliceous and other nodules which thus arise may reach several inches in diameter. Some of them have been supposed to represent lithophyses[2].

The Ordovician rhyolites of Westmorland[3] closely resemble the preceding, but in certain flows shew a very perfect microspherulitic structure. This is well seen in Long Sleddale[4] and near Great Yarlside (fig. 23). The large altered spherulites or nodules also occur. Good examples of these, as well as of devitrified obsidian with perlitic structure, are found also at Bouley Bay, in Jersey[5].

As a secondary alteration some of these Westmorland lavas shew silicification, in which much of the ground-mass and sometimes the porphyritic felspars are replaced by microcrystalline quartz. The same thing is seen in Caernarvonshire[6], and even more markedly in some ancient rocks, which have probably been rhyolites, at Trefgarn and Roche Castle, in Pembrokeshire.

Mr Allport was the first to give a clear account of some of the old altered volcanic glasses and to compare them with fresh Tertiary examples. He described what seems to be a devitrified and altered spherulitic rhyolite of pre-Cambrian age from Overley Hill or the Lea Rock near Wellington, Shropshire[7]. A few phenocrysts occur, but the bulk of the rock has been a glass enclosing numerous bands of spherulites. The glass is now devitrified, but perlitic cracks, marked by

[1] *Bala Volc. Ser. of Caern.* 35–39.
[2] Cole, *Q. J. G. S.* (1892) xlviii, 443–445, and references.
[3] *Q. J. G. S.* (1891) xlvii, 303.
[4] Rutley, *Q. J. G. S.* (1884) xl, Pl. XVIII, fig. 6, and Teall, pl. XXXVIII [1921].
[5] Davies, *M. M.* iii, 118, 119.
[6] Miss Raisin, *Q. J. G. S.* (1889) xlv, 253, 254.
[7] *Q. J. G. S.* (1877) xxxiii, 449–460; Teall, Pl. XXXIV, figs. 1, 2.

secondary products, are still evident. The spherulites too are for the most part much altered and stained red by iron-oxide.

Rhyolites presenting many features of interest occur in the Ordovician of Fishguard, in Pembrokeshire[1]. Mr F. R. C. Reed, who has made a study of these rocks, has shewn me in some of them beautiful examples of porphyritic micro-pegmatite, recalling the lavas described in America by Prof. Iddings (see fig. 34, *B*).

The old rhyolites of the Malvern Hills shew some curious features. Specimens from the New Reservoir, Malvern, are essentially cryptocrystalline rocks (perhaps devitrified), sometimes enclosing scattered phenocrysts of oligoclase. Narrow veins are occupied in some cases by infiltrated calcite, in others by a clear mosaic of quartz, orthoclase, and plagioclase of secondary formation. Mr Rutley[2] has described examples from the Herefordshire Beacon, in which old perlitic cracks are marked out by secondary epidote. The chief alteration-products are epidote and quartz, and the author suggests that some of the so-called epidosites (quartz-epidote-rocks) may have originated in this way.

Ancient acid lavas of Palæozoic and pre-Palæozoic ages occupy large tracts in the east of Canada and the United States. In spite of alteration, they have preserved many relics of original characteristic structures[3].

Among European examples of ancient rhyolites, Sauer has given an interesting description of the 'Meissen pitchstones' in Saxony. These are old rhyolites, probably of late Palæozoic age. They have been for the most part entirely glassy, often with beautiful perlitic fissuring, but some were spherulitic. The glassy character is in great measure retained, but there are often little patches[4] of cryptocrystalline or 'microfelsitic' nature or streaks of the same following

[1] Reed, *Q. J. G. S.* (1895) li, Pl. VI, figs. 3–5.
[2] *Q. J. G. S.* (1888) xliv, 740–744, Pl. XVII.
[3] G. H. Williams, *Amer. Journ. Sci.* (1892) xliv, 482–496; *Journ. of Geol.* (1894) ii, 1–31; Miss Bascom, *ibid.* (1893) i, 813–832.
[4] Rosenbusch-Iddings, Pl. II, fig. 5.

perlitic cracks and flow-lines. The rocks may become wholly devitrified in this fashion, and there is even a micro-crystalline type (the 'Dobritz porphyry'), representing an extreme alteration of obsidian by secondary devitrification[1]. The 'Fréjus pitchstones' in the Riviera are of similar character, but one variety ('pyroméride of Gargalong') is spherulitic on a relatively large scale.

The fresh Recent rhyolites of Pantellaria, an outlying island of the Lipari group, offer a peculiar type, rich in soda and iron (*pantellarite*). The phenocrysts are of anorthoclase and soda-sanidine, a green pleochroic augite, and the deep-brown, intensely pleochroic cossyrite. The ground-mass varies from almost holocrystalline to almost vitreous, a prevalent variety being a glass crowded with microlites of the above-mentioned minerals.

[1] For a brief notice of these rocks and also of the Devonian rhyolites of the Lenne district see *Sci. Progr.* (1894) ii, 54–56.

CHAPTER XII.

TRACHYTES AND PHONOLITES.

THE trachytes are lavas which, with a lower percentage of silica than the rhyolites, have as much or more of the alkalies. Consequently the typical *trachytes* consist essentially of alkali-felspars with a relatively small amount of coloured minerals and without free quartz. The name trachyte (given by Haüy to denote the rough aspect of the rocks in hand specimens) is used in the older literature to cover all the more acid half of the volcanic rocks. From it have been separated off, on the one hand, the rhyolites of modern nomenclature and, on the other, some hornblende- and mica-andesites, *etc.*

With the trachytes we shall treat some lavas of more peculiar constitution, in which a greater richness in alkalies has given rise to the formation of felspathoids as well as alkali-felspars: these are the *phonolites* and *leucitophyres.* The name phonolite (a translation of 'clinkstone,' from the supposed sonorous quality of the rock when struck) seems to have been in general use before the presence of microscopic nepheline in the rock was demonstrated, giving a character of precision to the definition. The original leucitophyres (of Coquand) were apparently any rocks with conspicuous crystals of leucite, but the name is now generally restricted to the type containing an alkali-felspar (sanidine) as an essential constituent.

Trachytes and phonolites exhibiting clearly their characteristic features are hitherto known chiefly among Tertiary and Recent volcanic products. It should be noticed, however, that these features are such as would readily be effaced in the older lavas. In any case, the Continental practice of restricting the family to Tertiary or later lavas rests on no philosophic ground, and indeed perfectly fresh trachytes and phonolites are known, *e.g.*, in the Lower Carboniferous of Scotland. The leucitophyres are a type of extremely restricted distribution, and the unstable nature of the characteristic mineral must make such rocks difficult to detect among the older lavas.

Constituent minerals. Felspars rich in potash or soda are by far the most abundant minerals in the rocks here considered. They occur both as phenocrysts and as the chief element in the ground-mass. The most prominent is usually orthoclase of the *sanidine* variety. In phenocrysts it has either a tabular or a columnar habit, and both may occur in the same rock. Carlsbad twinning is frequent, and in the larger crystals may shew the broken divisional line due to interpenetration. Some degree of zonary banding is sometimes found. The plagioclase felspar which occurs in many trachytes is usually *oligoclase*, but in more basic rocks we may find varieties richer in lime instead. The phenocrysts often shew carlsbad- as well as albite-twinning; zonary banding is not uncommon; and parallel intergrowth with sanidine may be noted (fig. 35).

In the true trachytes the most common ferro-magnesian element is perhaps brown *biotite*, in hexagonal flakes almost always affected by corrosion by the enclosing magma ('resorption'). This is shewn by a certain degree of rounding and the formation of a dark or opaque border, or even the total destruction of the flake, the resulting products being especially magnetite and sometimes greenish augite in minute granules. Brown *hornblende* is a less frequent constituent, in idiomorphic crystals with similar resorption-phenomena. The *augite*, which is scarcely less common than biotite as a constituent of trachytes, never shews this feature. It is usually pale green in thin slices. In the phonolites and leucitophyres the crystal often shews a deeper tint at the margin, and is almost always sensibly pleochroic, a character less common in the trachytes. Another pyroxene, *ægirine*, is characteristic of many phonolites and leucitophyres, but only occasionally present in the trachytes. It is green and pleochroic, with a much lower extinction-angle than the augites (5° or less in longitudinal sections). It sometimes grows round a kernel of augite with parallel orientation. The rhombic pyroxene of certain trachytes is always of a deeply coloured and vividly pleochroic variety (*hypersthene* or *amblystegite*), giving red-brown, yellow-brown, and green colours for the several principal directions of absorption.

The *nepheline* of the phonolites and leucitophyres occurs in minute crystals in the ground-mass, having the form of a short hexagonal prism with basal planes, and giving squarish or hexagonal sections. Owing to the small size of the crystals and the optical properties of the mineral, it is liable to be overlooked. Its decomposition gives rise to various soda-zeolites, which occur in nests and veins in many phonolites. The *leucite* of the leucitophyres is always idiomorphic, giving characteristic octagonal and rounded sections. Twin-lamellation is very frequent in the phenocrysts, but the smaller crystals which may occur often behave almost as if isotropic. The leucite may enclose needles of augite and crystals of the earlier-formed minerals, but not of felspar. Minerals of the sodalite-group are found in certain trachytes and constantly in the phonolites and leucitophyres. They are almost always in idiomorphic dodecahedra. The *sodalite* is clear when fresh, but often turbid from alteration : zonary structure is frequent. The blue *haüyne* is less often met with, but *nosean* may be very plentiful, usually forming crystals of some size, and always shewing more or less plainly its characteristic structure and border[1]. The sodalite-minerals give rise by alteration to natrolite and other zeolites.

Iron-ores (*magnetite*) occur but sparingly in these rocks. Yellowish *sphene* in good crystals is highly characteristic ; and *apatite* is common in colourless needles or sometimes in rather stouter prisms with violet dichroism. The trachytes usually contain a little *zircon* in minute prisms.

[1] Teall, Pl. XLI, fig. 1; XLVII, fig. 4.

Among less common minerals may be mentioned the *tridymite* of certain trachytes, in aggregates of minute crystals; *olivine*, as a rare constituent except in certain basic trachytes; and *melanite* garnet, which is found in some of the leucitophyres and certain phonolites in brown isotropic crystals belonging to an early stage of consolidation, sometimes shewing marked zonary banding[1].

As secondary products in trachytic, and also in andesitic, rocks, *opal* and other forms of soluble silica are not uncommon. Normally isotropic, these substances sometimes shew double refraction as a consequence of strain, usually about centres, so as to imitate a spherulitic structure. Opal sometimes encloses little flakes or aggregates of tridymite, or is coloured red by included scales of hæmatite. It occurs in the form of veins and irregular knots or patches.

Ground-mass. In contrast with the rhyolites, the rocks under consideration have few glassy representatives, and the ground-mass is frequently holocrystalline or at least with no sensible amount of glassy residue. This is especially true of the typical trachytes, which, with a chemical composition not very different from that of a mixture of felspars, have a strong tendency to crystallize bodily. Fluxional phenomena are not conspicuous, and the characteristic banding of the rhyolites is here wanting. Vesicular structure is rare, and perlitic cracks are not formed; but, in consequence of the crystalline nature of the ground, with a tendency to idiomorphism in its elements, a miarolitic or drusy structure may be met with. Any structure comparable with the spherulitic is uncommon, though a rough radial grouping of felspar prisms is sometimes observable.

Excluding the nepheline of the phonolites, non-felspathic constituents play in most cases a small part in the ground-mass of the rocks here considered. The ground consists, in the trachytes proper, essentially of minute felspars, which may, however, vary somewhat in habit. Most commonly they are 'lath-shaped' microlites, with some degree of parallel dis-

[1] Rosenbusch-Iddings, Pl. v, fig. 4.

position in consequence of flow, and this type of ground is so characteristic of these rocks that it is often styled the *trachytic*. On the other hand the minute felspars may have a shorter and stouter shape, recalling some of the rocks grouped above under the porphyries, and this structure is accordingly designated by Rosenbusch the *orthophyric*.

Phonolites poor in nepheline do not differ essentially in structures from the trachytes, but when the characteristic mineral is plentiful, forming very numerous minute crystals in the ground-mass, the general aspect of the latter is somewhat altered. The leucitophyres shew in their very variable structures further departures from the trachyte type; but all the rocks included in the present family resemble one another in being normally holocrystalline.

Leading types. Among the best known foreign *trachytes* are those of the Siebengebirge (Drachenfels type). Here a ground-mass of lath-shaped felspar microlites, with typical trachytic structure, encloses crystals of sanidine and oligoclase. The former are frequently of large size, and may shew carlsbad twinning. Biotite and magnetite occur sparingly. The rock of Perlenhardt, in the same district, exemplifies the orthophyric type of ground-mass of Rosenbusch. A little green augite accompanies the biotite, sphene is common, and sodalite occurs in crystals or crystalline patches. Trachytes from Solfatara and Mte Olibano near Naples shew similar characters.

Some very fresh augite-bearing trachytes occur as lava-flows and volcanic necks of Lower Carboniferous age in the Garlton Hills, Haddingtonshire[1]. These rocks consist of alkali-felspars with more or less of a bright to pale green, pleochroic augite, doubtless a soda-bearing variety. Specimens from Peppercraig (fig. 35) shew phenocrysts of sanidine, sometimes with intergrowths of oligoclase, in a holocrystalline ground-mass. The latter is chiefly of sanidine prisms, with a minor proportion of striated felspar. Augite builds imperfect crystals

[1] Hatch, *Trans. Roy. Soc. Edin.* (1892) xxxvii, 115–126; see also Geikie, *ibid.* (1879) xxix, Pl. XII, figs. 1, 2. The Carboniferous trachytes described by McMahon from Dartmoor seem to be much altered and their characters obscured.

and grains and numerous smaller granules; magnetite occurs sparingly in the same manner; and occasional needles of apatite are seen. In the type which forms the volcanic necks

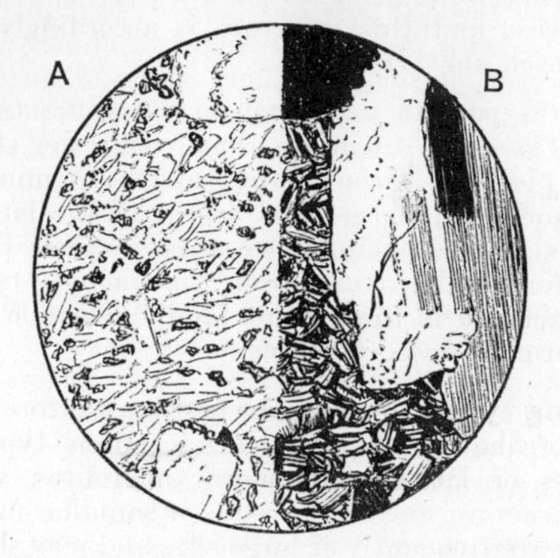

FIG. 35. AUGITE-TRACHYTE, PEPPERCRAIG, HADDINGTON; ×20. *A* in natural light, *B* between crossed nicols. Large phenocrysts of felspar are enclosed in a ground composed entirely of little felspar prisms and granules of augite [1980].

of North Berwick Law and the Bass Rock the porphyritic character is wanting, and the rock is of rather coarser texture, though otherwise comparable with the preceding. This type closely resembles some of the rocks named bostonite.

Examples of tridymite-bearing trachytes are afforded by some of the so-called 'domites' of Auvergne.

In those trachytes which in some respects approach the andesites, the coloured constituents, especially pyroxene, become relatively abundant, and plagioclase begins to predominate over orthoclase among the phenocrysts. A type from Mte Amiata in Tuscany and M. Dore in Auvergne contains a vividly pleochroic rhombic pyroxene (amblystegite) with subordinate biotite. Garnet and tridymite are accessories. The ground-mass of these rocks is of very variable character even in the same flow. Similar trachytes occur at Mocsár in Hungary, where the ground is sometimes largely glassy.

Purely glassy varieties (trachyte-obsidian) are uncommon in this family. In the localities where they are found, they are associated with trachytes wholly or mainly crystalline, or even narrow alternating bands occur of pure glass and of trachyte largely microcrystalline. Good examples of this occur in the Peak of Tenerife. It may be noted that a glassy variety of phonolite is also found in the Canaries, usually as a slaggy crust on the surface of a lava-flow. It is a brown or yellow glass with little development of crystallites.

The Arso-type of trachyte, the Ischia lava of A.D. 1302, approximates in some features to the basalts. The phenocrysts include, in addition to sanidine and a plagioclase felspar, abundant augite and olivine. The ground-mass is of felspar microlites with insterstitial glass, and is sometimes vesicular. Olivine-bearing trachytes occur also in the Azores.

Other trachytes shew an approach to the characters of phonolites in the abundance of sodalite, the occurrence of ægirine, *etc.* The trachytes of the Laacher See in the Eifel have crystals of sodalite and haüyne, besides sanidine and oligoclase. Biotite, brown hornblende, ægirine, sphene, magnetite, *etc.*, also occur, and the ground-mass is of the trachytic type. At the Laach volcano are also found ejected blocks of a rock named sanidine-trachyte or sanidinite. This consists essentially of sanidine with subordinate oligoclase, sodalite, occasional biotite, *etc.* Stellate groupings of crystals occur in both felspars, but on the whole the structure is that of a plutonic rather than a volcanic rock.

While the dominant mineral of the trachytic lavas is commonly a potash-felspar, there are some types very rich in soda, albite, anorthoclase, or some allied felspar occurring almost to the exclusion of sanidine or orthoclase. The 'quartzless pantellarites' of Pantellaria must be placed here, and the older equivalents of such types are to be sought among some of the rocks which have been styled quartzless ceratophyres. A very interesting soda-felspar-rock has been described from Dinas Head on the north coast of Cornwall[1]. This is probably to be regarded as an ancient lava, and it consists almost wholly of albite. Besides a compact variety, there are others which are spherulitic and nodular. The centre of a spherule is cryptocrystalline while its outer portion consists of radiating blades of albite. Such rocks may be termed old *soda-trachytes*, corresponding to the soda-rhyolites which are also known in this country.

At Traprain Law in the Garlton Hills, associated with the trachytes, occurs a *phonolite*[2]. It consists essentially of a mass of little sanidine prisms with a fluxional arrangement, in which lie ragged crystals of a bright green soda-augite. Small colourless patches are found on very close examination to consist of little crystals of nepheline with zeolitic decomposition-products. The mineral occurs only sparingly, and the rock thus belongs to that type of phonolite most nearly allied to the trachytes, with which its structure closely corresponds. Such rocks, the 'trachytoid' phonolites of Rosenbusch, are not the most characteristic type; and the 'nephelinitoid' group, in which the special mineral of the phonolites is more abundantly present, is commoner. Some of the Saxon phonolites are of the trachytoid type (Olbersdorf, near Zittau).

Of the commoner type of phonolite good examples occur in Bohemia (Brux, Teplitz, Marienberg, *etc.*), sanidine, nepheline, and ægirine being the essential minerals. Some varieties have conspicuous phenocrysts of sanidine. At the Roche Sanadoire[3] in Auvergne the porphyritic sanidines have often a core of plagioclase with parallel intergrowth, and little lath-shaped crystals of plagioclase occur also in the ground-mass.

Our only other British phonolite—that of the Wolf Rock[4] off the coast of Cornwall—is another good example. It belongs to the nosean-phonolites of some authors, that mineral being found plentifully in it, in addition to nepheline. The nosean occurs chiefly as phenocrysts with a dark interior and

[1] Howard Fox, *G. M.* 1895, 13–20.
[2] Hatch, *G. M.* 1892, 149, and *Tr. Roy. Soc. Edin.* (1892) xxxvii, p. 124.
[3] Fouqué and Lévy, Pl. XLVII, fig. 1; cf. fig. 2 and Pl. XLVI.
[4] Allport, *G. M.* 1871, 247–250; 1874, 462, 463; Teall, Pl. XLI, fig. 1.

clear border[1]. Sanidine is also found as phenocrysts. The general mass of the rock consists of lath-shaped sanidine crystals, more or less idiomorphic crystals of nepheline, and

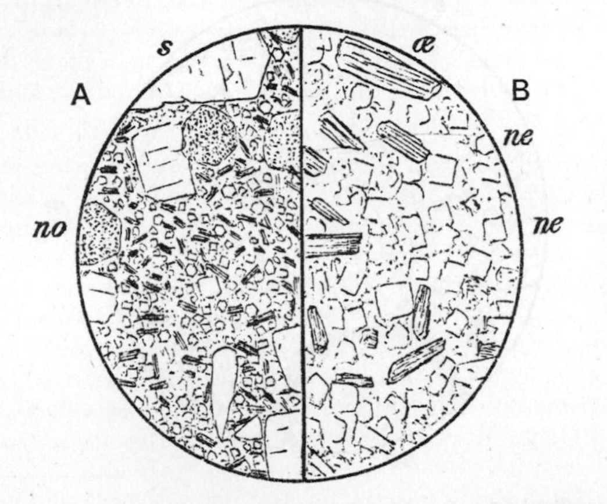

FIG. 36. PHONOLITE, WOLF ROCK, CORNWALL; $A \times 20$, $B \times 100$.

The figure shews phenocrysts of sanidine (*s*) and nosean (*no*) in a ground-mass of sanidine, nepheline (*ne*), and ægirine (*æ*) [1771].

little dirty green microlites of ægirine. Iron-ores are scarcely represented, and there is little or no residual glass (fig. 36).

The *leucitophyres* are a very small group of rocks known only from three or four districts and best developed in the late Tertiary lavas of the Eifel. The leucite is often of two generations, the larger crystals being frequently of irregular shape. It is always accompanied by nosean and sanidine (fig. 37). The ferro-magnesian mineral is a green pleochroic augite with zonary banding: the other constituents are sanidine, sphene, occasionally biotite, and often a little melanite. The structure of the rocks is very variable[2]. In some there is a well-

[1] Teall, Pl. XLVII, fig. 4 (misplaced 5 in key-plate).

[2] A. Martin, *M. M.* (1891) ix, 251 (*Abstr.*). For figures of leucitophyres see Fouqué and Lévy, Pl. XLVIII, fig. 1, and LI, fig. 1; Teall, Pl. XLI, fig. 2, and XLVII, fig. 4; Rosenbusch-Iddings, Pl. V, fig. 2; XV, fig. 1; XVI, fig. 6; XVII, fig. 1.

defined ground-mass of minute nepheline, sanidine, augite, and leucite, enclosing phenocrysts of leucite and nosean (Olbrück,

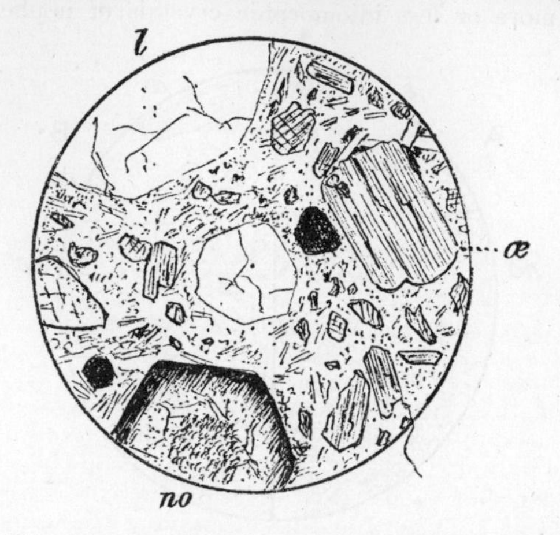

FIG. 37. LEUCITOPHYRE, RIEDEN, EIFEL; ×20.

The larger elements shewn are leucite (*l*), nosean (*no*), and ægirine (*æ*) [160].

etc.). In other varieties there is but little sanidine (Schorenberg), while others again have sanidine in large shapeless plates enclosing the other constituents instead of a ground-mass (Perlerkopf). Pumiceous modifications of leucitophyre occur, especially as fragments in the associated tuffs.

CHAPTER XIII.

ANDESITES.

UNDER this family we include all the lavas of 'intermediate' composition not embraced in the preceding family. The name andesite, first used by von Buch and derived from the prevalence of such rocks in the Andes, is roughly equivalent to Abich's 'trachydolerite,' implying the intermediate position of these lavas between the acid ones (trachytes of older writers) and the basic (dolerites). The characteristic minerals are a soda-lime-felspar and one or more ferro-magnesian minerals. The alkali-felspars and quartz of the acid rocks are typically absent, as arc also the lime-felspar and olivine of the basic rocks. The andesites are distinguished, according to the dominant ferro-magnesian constituent, as *hornblende-*, *mica-*, *augite-*, and *hypersthene-andesites.* Further there is usually recognised a quartz-bearing and more acid division, known as *dacites* or quartz-andesites. Having regard to true lavas, these quartz-bearing andesites seem to be of somewhat limited distribution: many of the rocks described as 'dacites' are of intrusive types, and belong to the less acid quartz-porphyries.

Those petrologists who restrict the name andesite to rocks of late geological age, apply to their pre-Tertiary equivalents the name 'porphyrite[1].' Under the same title they include

[1] Many of the rocks designated 'melaphyre' are pyroxene-andesites, others being basalts.

various rocks of intrusive types, and it is to these latter that we have already confined the name. Again, certain English petrologists have used the name porphyrite for andesites which have undergone some degree of change by weathering, *etc.*, a distinction which seems scarcely important enough to be recognized in classification or nomenclature.

As regards the general affinities of the family, the dacites have features in common with the rhyolites, the hornblende- and mica-andesites with the trachytes, and the pyroxene-andesites with the basalts, marking thus the intermediate position held among the volcanic rocks by the lavas here considered. As regards the appropriateness of the name, it is remarkable that the lavas of the great volcanic belt of the Andes belong, in so far as they are known, almost exclusively to this family[1].

Phenocrysts. Soda-lime-felspars are the most abundant elements porphyritically developed in these rocks. They include members varying from oligoclase to bytownite, but *andesine* and *labradorite* are the most common. As a rule the more acid plagioclase belongs to the hornblende- and mica-andesites and dacites, the more basic to the pyroxene-andesites[2]. The crystals, however, are often strongly zoned, shewing a change from a more basic variety in the centre to a more acid at the margin. They are idiomorphic and of tabular habit. With albite-lamellation is frequently associated twinning on the pericline or on the carlsbad law. The commonest inclusions are glass-cavities[3], either as 'negative crystals,' or rounded: sometimes large irregular cavities occupy much of the bulk of a crystal. The decomposition-products of the felspars are calcite, finely divided kaolin or mica, epidote, quartz, *etc.* When an alkali-felspar occurs as an accessory, it has the same characters as in the rhyolites and trachytes.

[1] *Cf.* Iddings, *Journ. of Geol.* (1893) i, 164–175.

[2] French petrologists recognize 'andesites' and 'labradorites' as distinct rock-types, characterized by andesine and labrador felspar respectively, but this is with reference to the ground-mass.

[3] See Zirkel, *Micro. Petr. Fortieth Parall.*, Pl. v, fig. 3; xi, fig. 2.

The *hornblende* of andesites is in idiomorphic prisms, often twinned. It is usually a brown pleochroic variety with quite low extinction-angle, but green hornblende also occurs. The mica is a brown, strongly pleochroic *biotite* with extinction sometimes sufficiently oblique to shew lamellar twinning parallel to the base. Both hornblende and biotite shew the same resorption-phenomena[1] as in the trachytes. Their decomposition-products are chlorite, magnetite, carbonates, *etc.*

The *augite* is in well-shaped crystals, light green and usually without sensible pleochroism. Twin-lamellation is common. Alteration may give rise to chlorite, epidote, calcite, *etc.* The rhombic pyroxene in the andesites is usually *hypersthene*[2] or at least a distinctly coloured and more or less pleochroic variety. It builds idiomorphic crystals in which the pinacoid faces are more developed than the prism; so that the cross-section is a square with truncated corners as contrasted with the regular octagon of augite. In longitudinal sections the straight extinction is of course characteristic. The rhombic pyroxene is often converted in the older rocks to bastite.

The *quartz* of the dacites is either in good hexagonal pyramids or more or less rounded and corroded, with inlets of the ground-mass.

Original iron-ores are usually not abundant: *magnetite* is the only one commonly found. Needles of *apatite* occur, and in the more acid andesites little *zircons.* Some of the more basic rocks have sparingly phenocrysts of *olivine.* As occasional accessories may be noted *tridymite*[3] (in druses), *garnet,* and *cordierite*[4].

Structure of ground-mass. In many andesites the only mineral which occurs distinctly in two generations is

[1] Fouqué and Michel Lévy, Pl. XXVIII, XXIX; Zirkel, *Micro. Petr. Fortieth Parallel,* Pl. V, fig. 2.

[2] See Whitman Cross, '*On Hypersthene-Andesite...*' (1883) *Bull. No.* 1 *U.S. Geol. Surv.*; *Amer. Journ. Sci.* (1883) xxv, 139; Teall, *G. M.* 1883, 145–148.

[3] Koto, *Q. J. G. S.* (1884) xl, 441, 444.

[4] Osann, *M. M.* viii, 284–285 (*Abstr.*).

the felspar. The felspar of the ground-mass builds little 'lath-shaped' crystals, often simple, sometimes twinned but usually without repetition. It is probably, as a rule, of a more acid variety than the porphyritic felspar, andesine or oligoclase occurring in different cases. Augite may also be present as a constituent of the ground-mass, forming very small crystals of pale-green tint.

Some of the hornblende- and mica-andesites have a *trachytic* type of ground-mass, composed essentially of very small felspar-laths with little or no glassy base, as in the Drachenfels trachyte. It is not always easy to ascertain whether any glass is present or not. From this type, as from the others, there are, however, transitions to rocks with a ground-mass mainly glassy.

Less common is a '*microfelsitic*' or cryptocrystalline structure. This is seen in some of the dacites. In some cases spherulitic structures are found.

In most typical andesites, and especially in the pyroxene-bearing kinds, the ground-mass has the very distinctive 'felted' character termed by Rosenbusch *hyalopilitic*. This consists of innumerable small felspar-laths, simple or once twinned, often with evident flow-structure, and a residuum of glassy matter. Vesicles are common, and their infilling by secondary products gives rise to amygdules[1]. So characteristic is this type, that it is often spoken of as the 'andesitic' ground-mass. When the little felspars are closely packed together, to the exclusion of any glassy base, we have the *pilotaxitic* structure of Rosenbusch. On the other hand, by increase in the proportion of isotropic base, these andesites graduate into more or less perfectly *glassy* forms. Wholly glassy types (andesite-obsidian, including andesite-pumice), are known in small development only, except in so far as they form part of tuffs, *etc.* Some andesitic rocks shew various kinds of *variolitic* structures[2] comparable with those seen in basalts (see fig. 41 *A*, on p. 171).

[1] Very many of the amygdaloidal lavas (Ger. Mandelstein) belong here.

[2] *G. M.* 1894, 551–553 (Carrock Fell dykes).

Leading types. Of fresh *andesite-glasses* the best known are those of Santorin[1], reproducing the perlitic fissures and other features of the acid obsidians. These seem to be in the main hornblende-bearing, but contain augite associated with that mineral. Vesicular and pumiceous modifications are found.

In Scotland andesitic rocks of Tertiary, and some of Palæozoic, age have more or less glassy modifications, but these are only of local occurrence.

The interesting rock described by Prof. Judd[2] from the (probably) Old Red Sandstone breccia near Scroggieside Farm in N.E. Fife is on the border-line between rhyolite and dacite. It has a glassy modification which the author styles mica-dacite-glass. Phenocrysts of oligoclase and deep brown biotite are imbedded in a glassy ground-mass containing trichites, globulites, and imperfect microlites of felspar (perhaps orthoclase). The glass shews beautiful perlitic fissures.

Little is known of true *dacites* among the Lower Palæozoic lavas of this country, though some of the rocks included above as rhyolites would probably be styled dacites by certain petrologists. The name has also, as remarked above, been applied to some of the acid intrusives.

Among foreign dacites the best known are those of Tertiary age in Transylvania[3] and Hungary and in some parts of the Andes. They include holocrystalline examples (Kis Sebes, Rodna) and others with cryptocrystalline and microspherulitic ground-mass (Schemnitz district, *etc.*), as well as those having the hyalopilitic structure so common among the andesites. Hornblende-dacite with microspherulitic structure occurs among the recent lavas of Santorin in the Grecian Archipelago[4]. Zeolites and isotropic opal are found as secondary products, or in other cases chalcedony[5].

[1] Fouqué and Lévy, Pl. xxx ; xxxi, fig. 1.

[2] *Q. J. G. S.* (1886) xlii, 427–429, Pl. xiii, figs. 7, 8.

[3] The name was first used by Stache for quartz-bearing andesites in Transylvania (Dacia).

[4] Fouqué and Lévy, *Min. Micr.*, Pl. xviii.

[5] *Ibid.* Pl. xvii, fig. 2.

The andesites characterized by biotite or hornblende have affinities, as already remarked, with the typical trachytes. A *mica-andesite* free from hornblende is exceptional, but the name may be applied to varieties in which biotite is the dominant, though not the sole, ferro-magnesian constituent. Examples occur in the Schemnitz district of Hungary (Giesshübel, *etc.*). The rocks usually taken as the type of *hornblende-andesite* are those of the Tertiary volcanic district of the Siebengebirge, near Bonn, already alluded to as the home of certain typical trachytes. In addition to abundant brown

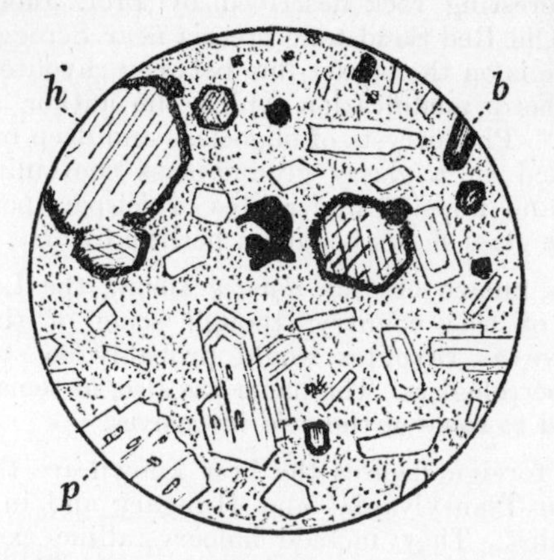

FIG. 38. HORNBLENDE-ANDESITE, STENZELBERG, SIEBENGEBIRGE; ×20.

The hornblende (*h*) and subordinate biotite (*b*) shew the resorption-border; the phenocrysts of felspar (*p*) shew zonary banding and glass-cavities; the ground-mass is only diagrammatically represented [117 *a*].

hornblende, these andesites contain more or less biotite and a few prisms or grains of pale green augite. The two former minerals always shew the phenomenon of resorption. The Bolvershahn rock, with a considerable amount of deep brown biotite, may be called a hornblende-mica-andesite. The felspar phenocrysts shew very marked zonary banding in polarized light. The Wolkenburg rock is a characteristic hornblende-andesite. Its phenocrysts include the three ferro-magnesian minerals mentioned, hornblende largely predominating, good crystals of andesine, and a little magnetite and apatite, while its ground-mass is of the trachytic type. A very similar rock is that of Stenzelberg (fig. 38), in which some of the hornblende crystals attain a conspicuous size.

The late Tertiary lavas of France include representatives of this group, *e.g.* near Lioran in the department of Cantal. Fouqué and Michel Lévy[1] describe a mica-andesite from that neighbourhood, having biotite as its prominent coloured constituent, while subordinate hornblende and augite also occur, and the felspar phenocrysts are of oligoclase. Tridymite is found in this rock. Other examples are hornblende-andesites with conspicuous phenocrysts of brown hornblende, together with pale augite, zoned felspars, and complex groupings of magnetite, in a trachytic ground-mass with fluxion-structure.

Numerous hornblende- and mica-andesites have been described by Zirkel, Iddings, and others in the Western States of America. In our own country these rocks are very poorly represented. One good example occurs on the summit of Beinn Nevis[2], and, though probably of Carboniferous age, it is fairly fresh. The phenocrysts are of light-brown idiomorphic hornblende and of plagioclase full of glass-inclusions, *etc.* The ground-mass is obscured by specks of iron-ore and alteration-products, but is seen to consist largely of densely packed, minute felspar-microlites.

Andesites having a pyroxene as their dominant non-felspathic constituent are perhaps more widely distributed than any other group of lavas, and are largely represented among the products of volcanoes now active. Since a rhombic and a monoclinic pyroxene are often associated, the rocks are spoken of as *pyroxene-andesites*, while the marked predominance of one or other of these minerals gives a hypersthene- or an augite-andesite. The two pyroxenes may be distinguished by the characters noted above, the direction of extinction being always observed. Pleochroism is not a conclusive test, since it may be inappreciable in some small

[1] *Min. Micro.* Pl. XXII, XXVIII. An andesite with hornblende subordinate to augite is figured in Pl. XXIX. See also Pl. XXXVIII.

[2] Teall, Pl. XXXVII, fig. 1.

crystals of the rhombic mineral. On the other hand Koto[1] notes distinctly pleochroic augite in some Japanese augite-andesites. The rocks he describes are rather peculiar, being mostly holocrystalline and sometimes rich in tridymite.

Hypersthene-andesites[2], or hypersthene-augite-andesites in which the rhombic pyroxene predominates over the monoclinic, are especially widely distributed among the lavas of different periods. Prof. Judd[3] has pointed out that the same general petrographical type is found in lavas ranging in chemical composition from basalt to dacite. Thus the basic dykes of Santorin, the lava of Buffalo Peaks in Colorado[4], the Cheviot rocks, the recent lavas of Santorin, and the rocks of Krakatau consist of the same minerals in a glassy base of the same general composition, but the relative proportions of the minerals (in the aggregate basic) to glass (decidedly acid) varies in the different cases from 9 : 1 to 1 : 9. This illustrates the impossibility of naturally classifying by mineralogical characters alone rocks which have a glassy base.

The Ordovician volcanic series of the Lake District includes a considerable development of pyroxene-andesites, some of which contain plenty of pseudomorphs after a rhombic pyroxene (Falcon Crag near Keswick, *etc.*), while many others are characterized by monoclinic pyroxene only. A few of these rocks have been described by Mr Clifton Ward, Prof. Bonney, and Mr Hutchings[5]. The ground-mass is usually typically hyalopilitic.

In the Bala series of Caernarvonshire there are few andesites. Some, with augite only, occur in the Lleyn district[6], and one with dominant hypersthene forms an intrusive mass

[1] *Q. J. G. S.* (1884) xl, 431–447.

[2] See Whitman Cross, *cit. sub.* For a detailed description of fresh Tertiary lavas of this type, as well as of hornblende-andesites and dacites, see Iddings, *Geology of the Eureka District*, *Monog.* xx *U.S. Geol. Surv.* (1893) and plates.

[3] *G. M.* 1888, 1–11.

[4] This is the lava in which the existence of a rhombic pyroxene was first verified. Whitman Cross, *Bull. No.* 1 *U.S. Geol. Surv.* and *Amer. Journ. Sci.* (1883) xxv, 139.

[5] *G. M.* 1891, 539–544.

[6] *Bala Volc. Ser. Caern.* 68.

at Carn Boduan[1] in the same district (fig. 39). No detailed study has yet been made of the Arenig pyroxene-andesites of Merioneth, but lavas of the same approximate age in the

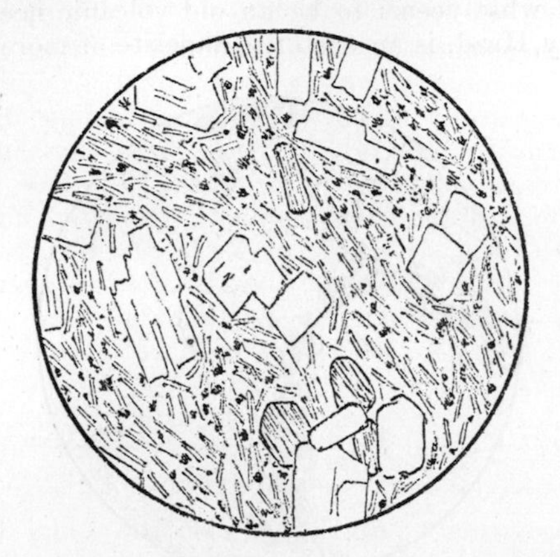

FIG. 39. HYPERSTHENE-ANDESITE, CARN BODUAN, CAERNARVONSHIRE; ×20.

Phenocrysts of felspar and bastite pseudomorphs after hypersthene are set in a ground-mass, in which felspar-microlites with partial fluxional arrangement are the most conspicuous element [643].

Stapeley Hills (Todleth, *etc.*) in Shropshire are of the same general type as the Cheviot rocks, containing both rhombic and monoclinic pyroxenes, and this is true also of the Bala lavas of the Breidden Hills (Moel-y-golfa, *etc.*)[2].

Many of the old lavas loosely grouped under the field-term 'porphyrite' in the Old Red Sandstone and Carboniferous of Scotland are andesites, ranging in composition from a relatively acid type (dacite) to varieties verging on basalt. One of the former, from North-east Fife, has already been mentioned. In the same district are good examples of more basic types also[3]. One quarried at Northfield is an augite-

[1] *Bala Volc. Ser. Caern.* 69–71.

[2] Watts, *Q. J. G. S.* (1885) xli, 539–543; *Proc. Geol. Assoc.* (1894) xiii, 337–339, with figures.

[3] Judd, *Q. J. G. S.* (1886) xlii, 425–427, Pl. XIII, figs. 1, 2.

andesite with phenocrysts of augite, and perhaps subordinate enstatite, in a ground-mass largely of glass filled with globulites and trichites and a profusion of felspar-microlites. Another, from what seems to be an old volcanic neck, quarried at Causeway Head, is an enstatite-andesite of more crystalline

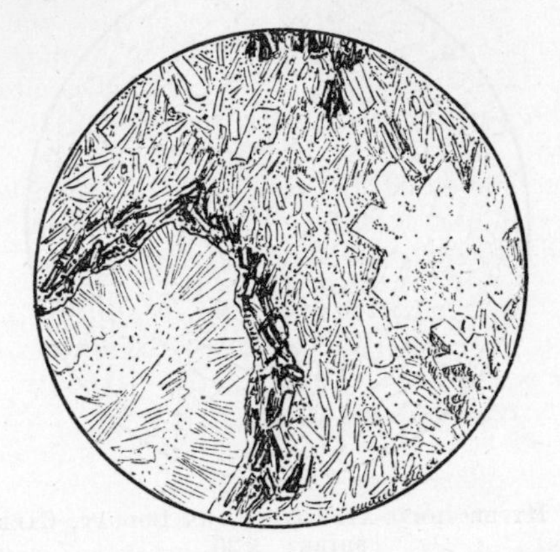

FIG. 40. AMYGDALOIDAL AUGITE-ANDESITE, STOCKDALE, WESTMORLAND; ×20.

The rock is considerably weathered, the augite being wholly replaced by a green chloritoid mineral. The same substance, in bunches of little scales, lines the vesicle seen on the left of the figure, and a ring of opaque decomposition products borders the vesicle. On the right is a group of felspar phenocrysts [758].

type, and is an aggregate of little prisms of triclinic felspar (near andesine), prisms and granules of pale enstatite, and grains of magnetite, with very little glassy residue. There are no porphyritic elements. Subordinate augite accompanies the rhombic pyroxene, and biotite is another accessory.

The Old Red Sandstone lavas of the Cheviots[1] are mostly hypersthene-andesites, containing both rhombic and monoclinic pyroxenes. The freshest type shews phenocrysts of labradorite, honeycombed with inclusions of ground-mass,

[1] Teall, Pl. XXXVI, XXXVII, fig. 2; *G. M.* 1883, 102–106, 146–152, Pl. IV, 252–254. See also Petersen, *G. M.* 1884, 226–234 (*Abstr.*).

crystals of hypersthene shewing distinct pleochroism, and crystals and grains of pale augite, in a ground-mass of pale brown glass and felspar microlites. The ground often has flow-structure, and shews varieties of the hyalopilitic type. The iron-ores are represented by magnetite and minute red scales of hæmatite. The rock is often veined by opal or chalcedony, stained red with ferric oxide. The more weathered lavas of the district (part of the 'porphyrites' of some authors) have had similar characters, but the felspars and pyroxenes are more or less decomposed, and the ground obscured by ferruginous matter. There are sometimes vesicles, filled with chalcedony, *etc.*

Certain dykes described by Mr Teall[1] in the North of England may be referred to here, being petrographically augite-andesites. Some of them (Cleveland, *etc.*) must be of Tertiary age, while others are probably late Palæozoic. The Cleveland dyke is traced from near Whitby to Armathwaite near Carlisle, and perhaps farther. It contains porphyritic felspars, often broken, in a ground-mass composed of small felspar crystals, minute crystals and grains of augite, crystals of magnetite, and abundant interstitial matter. This last is sometimes glassy, but commonly charged with various products of devitrification, giving a decided reaction with polarized light. The Acklington dyke is similar, but usually without the porphyritic crystals and with less of the interstitial base.

The Tynemouth dyke is less fine-textured. It contains porphyritic aggregates of anorthite crystals in a ground-mass of elongated lath-shaped felspars, grains of augite, magnetite, and a considerable amount of interstitial base with devitrification-products and microlites and skeletons of felspar[2]. There are small spherical vesicles filled either by secondary minerals or by the base itself, which has broken in before its final consolidation[3]. The dykes of Hebburn, Brunton,

[1] *Q. J. G. S.* (1884) xl, 209–247, Pl. XII, XIII; *Brit. Petr.* Pl. XII, XIV.
[2] The structure is the 'intersertal' of Rosenbusch, who cites these dykes as examples of his type 'tholeiite.'
[3] Teall, *G. M.* 1889, 481–483, Pl. XIV.

Seaton, and Hartley are similar, though usually without the large felspars.

The interstitial base of these rocks, with its enclosed crystallitic bodies and its devitrification phenomena, presents various points of interest described in the papers cited. Mr Teall points out the resemblance of the Cleveland dyke to the Eskdale dyke in the south of Scotland, which has been described by Sir A. Geikie[1], and which in places consists very largely of glassy matter enclosing various crystallitic growths.

The Tertiary andesitic lavas of the Western Isles of Scotland, as described by Prof. Judd[2], are of somewhat peculiar character. There are more acid types with hornblende or biotite, or both, and less acid with pyroxenes, augite predominating. Structures approaching the holocrystalline (doleritic) are common, though other kinds are found in great variety, and some have been largely glassy. The most striking feature, however, is the wide-spread chemical alteration which has affected the rocks, obliterating to a great extent their original characters, and giving rise to abundant epidote, chlorite, and other secondary products, including sometimes pyrites. In their frequent dioritic or doleritic aspect and their peculiar mode of decomposition these lavas resemble very closely the rocks to which the name '*propylite*' has been applied in Hungary and the Western States of America. It is now generally recognized that the rocks to which this name was applied by Richthofen, Zirkel, and others, do not constitute a distinct family, but are altered forms, partly of andesites, partly of various intrusive rocks. This appears, for instance, from Zirkel's own account of the differences between the 'propylites' and andesites of the Western States[3]. The 'green hornblendes' supposed to characterize the former are, according to Wadsworth, chloritic pseudomorphs.

[1] *Proc. Roy. Phys. Soc. Edin.* (1880) v, 219. See also Teall, p. 196 and Pl. XXIV, fig. 1.

[2] *Q. J. G. S.* (1890) xlvi, 341–382.

[3] *Micro. Petro. Fortieth Parallel*, 132, 133.

CHAPTER XIV.

BASALTS.

In the *basalt* family we include all the basic lavas except those in which a relatively high content of alkalies has given rise to the formation of minerals of the felspathoid group. The rocks range in texture from vitreous to holocrystalline. Except in a few of the latter (*dolerites*), the distinction between phenocrysts and ground-mass is commonly well marked, but the relative proportions of the two vary greatly in different types. The characteristic minerals in this family of rocks are a felspar rich in lime, augite, and olivine.

Following our principle, we shall make no distinction, as regards nomenclature and classification, between Tertiary and pre-Tertiary lavas. Foreign petrologists usually restrict the names basalt and dolerite to the newer examples, their older equivalents being denoted by such names as melaphyre, augite-porphyrite, diabase, *etc.*, some of which are also applied to rocks of the intrusive division.

Certain exceptional lavas (*limburgites*, *etc.*) which are of ultra basic, rather than normally basic, composition, will be briefly noticed. Some of them probably correspond rather to the nepheline-basalts, *etc.*, treated in the succeeding chapter.

Constituent minerals. The felspars of the basalts are of decidedly basic varieties. When distinctly porphyritic crystals occur, they seem to be usually *bytownite* or *anorthite*, while the felspars of the ground-mass are more commonly *labradorite.* The phenocrysts shew albite-lamellation, often combined with pericline- and carlsbad-twinning. Zonary structure and zonary arrangement of glass-cavities are met with. The felspars of the ground-mass have the lath-shape, and are commonly too narrow to shew repeated twinning.

The dominant pyroxenic constituent is an ordinary *augite*, and this too may occur in two generations. If so, the phenocrysts often have good crystal-forms, with octagonal cross-section, twinning is frequently seen, and sometimes zoning and hour-glass-structure. The colour is usually very pale, brownish or more rarely greenish, the latter especially in the interior of a crystal. The augite of the ground-mass is either in little idiomorphic prisms or in granules, and is often very abundant. Decomposition of the augite produces chloritoid substances, *etc.*[1] A rhombic pyroxene, *hypersthene* or bronzite, occurs only in certain basalts, where it seems to some extent to take the place of olivine. It is always in idiomorphic prisms, and in the older rocks is very generally serpentinized. Some basalts, again, contain corroded crystals of *brown hornblende*, and others a little *brown mica*.

Octahedra and grains of *magnetite* are generally abundant, and this mineral frequently recurs in a second generation in little granules. Besides this, there are frequently little opaque or deep brown scales of *ilmenite* or deep red flakes of *hæmatite*. Grains of *native iron* occur locally in a few basalts (Ovifak in Disco, Greenland)[2].

In by far the greater part of the basalts *olivine* is an essential constituent, and in many it is abundant, though confined, as a rule, to phenocrysts. These are sometimes well shaped crystals, sometimes more or less rounded. The mineral is colourless or very pale green. It often shews serpentine-strings following cleavage- or other cracks[3], and with further alteration passes into various secondary products, serpentine, carbonates, *etc.* Another common change is the production

[1] Teall, Pl. XXII, fig. 2.

[2] Fouqué and Lévy, *Min. Micr.* Pl. XXXVI, fig. 2; Steenstrup, *M. M.* i, 148, Pl. VI.

[3] Zirkel, *Micro. Petr. Fortieth Parallel*, Pl. X, fig. 3; XI, fig. 3.

of a red or brown margin to the olivine, due to iron-oxide, the olivine in basalts being often of a variety rich in iron.

Of other minerals we need only note *apatite*, which is not uncommon, forming long needles, either colourless or of a faint violet or bluish tint.

A peculiar feature in certain American basalts[1] is the occurrence of isolated grains of *quartz*. These are always corroded by the magma and usually surrounded by a ring of augite or its alteration-products. They do not seem to represent fragments mechanically enclosed by the lava, but have originated under special conditions at an early stage of its history. They are comparable with similar grains in many lamprophyres (see above, p. 122).

Structure. The rocks of the basalt family present a wide range of characters, from purely glassy examples at one extreme to wholly crystalline at the other. Rocks exhibiting such a range may occur, perhaps exceptionally, in one district, their petrological characters being correlated with their various modes of occurrence, as is well described by Prof. Judd[2]. On the whole, the tendency to crystallization is much stronger here than in the more acid families of lavas. Again, the order of crystallization of the several constituents is less strongly marked, the mutual relations between augite and felspar, in respect of priority, varying, while the iron-ores, though they commonly begin to crystallize at an early stage, may be in part rather late. These remarks are true of both the 'intratelluric' and the 'effusive' periods, when these are distinctly separable, but in some of the holocrystalline types the porphyritic character is not recognizable. Some of these rocks differ in no essential from those already described as diabases, the petrological distinction between the intrusive and the volcanic types not being marked by any hard and fast line.

[1] Diller, *Amer. Journ. Sci.* (1887) xxxiii, 45–49; also *Bull. No.* 79 *U.S. Geol. Surv., A Late Volcanic Eruption in N. California*, 24–29; Iddings, *Amer. Journ. Sci.* (1888) xxxvi, 209–213; *Geology of the Eureka District* (*Monog. U.S. Geol. Surv.*) 393, Pl. IV, fig. 4.

[2] *Q. J. G. S.* (1886) xlii, 66–82, Pl. V, VI.

Except in the form of lapilli and fragments in tuffs, the purely vitreous type, *tachylyte*, is of very limited distribution, being found only as a very thin crust on some lava-flows or a narrow selvage to basalt-dykes. It consists of a brown or yellow glass densely charged with a separation of magnetite. This is sometimes in globulites disseminated through the glass so as to render it almost opaque, or collected in cloudy patches (cumulites); at other times it forms trichites or crystallites of minute size[1]. Perlitic structure is less common than in the obsidians. Interesting spherulitic structures are met with in some examples[2]. When distinct phenocrysts occur abundantly in the glassy ground-mass, we have what is sometimes called the 'vitrophyric' structure. The basic glass is subject to secondary changes, probably involving, as a rule, hydration and other chemical changes, but the resulting substance, known as palagonite, is still an isotropic glass, yellow, brown, or sometimes green in sections.

Radiate aggregates of felspar microlites or fibres, answering to the spherulites of acid rocks, occur in some basaltic glasses, which are known as *variolites*. These aggregates vary in size and in the regularity of their structure, which ranges from mere fan-like and sheaf-like groupings (*cf.* fig. 41, *A*) to spherules with a perfect radiate structure. They may occur isolated in a glassy matrix, or coalesce into bands, or form a densely packed mass with little or no interstitial matter. The variolites are very susceptible to alteration.

Leaving the glassy basalts, we note those in which the ground-mass enclosing the phenocrysts of olivine, augite, felspar, *etc.*, is *hypocrystalline*, consisting of lath-shaped felspar-microlites and granules or microlites of augite with more or less of a residual glassy base. Of this division there are various types, depending on the relative proportions of augite, felspar, and glass, and the mutual relations of the minerals. When the felspar-microlites preponderate, usually with a more or less fluxional arrangement, the ground-mass does not differ essentially from the 'hyalopilitic' type so common in

[1] Judd and Cole, *Q. J. G. S.* (1883) xxxix, Pl. xiv.
[2] Cole, *ibid.* (1888) xliv, 300–307, Pl. xi.

the pyroxene-andesites. Vesicles are frequent in such rocks. More often, however, augite is abundantly represented in the basaltic ground-mass. Again, unindividualised glass may form

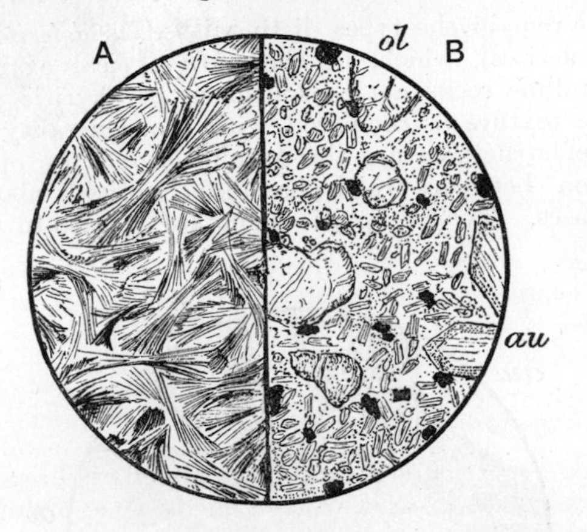

Fig. 41.

A. Andesite vein approaching the structure of variolite, Carrock Fell, Cumberland; ×20, crossed nicols. This is of the type which consists essentially of radiating felspar fibres grouped in sheaf-like bundles. There are also skeleton-prisms of a pyroxenic mineral, better seen in natural light [1552]. *B.* Limburgite, Whitelaw Hill, Haddington; ×20, natural light. Phenocrysts of olivine (*ol*), zoned augite (*au*), and magnetite are enclosed in a ground-mass of glass containing abundant prisms and granules of augite but no felspar. The glass, which constitutes the bulk of the ground, varies from brown to nearly colourless [1982].

the bulk of the ground, and this is especially the case in the limburgites (fig. 41, *B*).

By the failure of the glassy residue we pass to those types of basalt in which the phenocrysts are enclosed in a *holocrystalline* ground-mass. Here again there are numerous varieties. Sometimes little eye-like or lenticular patches relatively rich in augite are contrasted with adjacent patches rich in felspar. When felspar-microlites make up a large part of the ground-mass, we have a structure analogous to the 'pilotaxitic' of some andesites and trachytes, the flow being more or less marked. On the other hand, the ground may consist mainly of small rounded granules of augite, between which the little felspars seem to be squeezed.

There remain the types distinguished as *dolerites* (usually olivine-dolerites), which, in the most typical examples, are holocrystalline rocks not conspicuously porphyritic, sometimes of coarse texture as compared with the generality of lavas. The chief structures are the granulitic and the ophitic, the distinction between which has been noticed above under the diabases. Typical ophitic structure is rare in true lava-

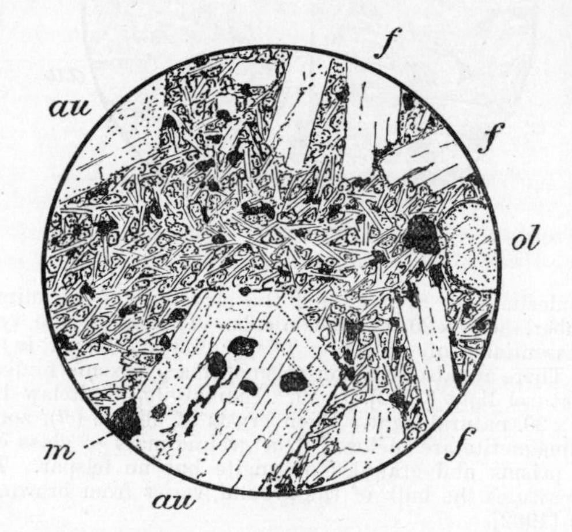

Fig. 42. Basalt, Etna lava of 1669 eruption, Catania; ×20: Shewing phenocrysts of augite (*au*), felspar (*f*), olivine (*ol*), and magnetite (*m*) in a holocrystalline ground-mass of little lath-shaped felspars and granules of augite and magnetite [131].

flows. The 'intersertal' structure of Rosenbusch corresponds in part with the granulitic, but it also includes the type in which some residual glass, as well as augite and other minerals, occurs in the interstices between the lath-shaped felspar crystals. Only exceptionally in doleritic lavas do we find an idiomorphic development of the augite and an approach in structural characters to some plutonic types (*e.g.* the Löwenburg olivine-dolerite in the Siebengebirge).

Some dolerites enclose large scattered porphyritic crystals of felspar. In other cases there are porphyritic aggregates of crystals (felspar, olivine, augite, *etc.*) having the mutual relations characteristic of plutonic rocks : this is the *glomero-porphyritic* structure of Prof. Judd[1]. It is not confined to the holocrystalline dolerites. The crystals forming such a hypidiomorphic aggregate may still present idiomorphic outlines towards the surrounding rock[2].

Many of the Tertiary basalts in Germany, *etc.*, enclose so-called '*olivine-nodules*,' which are hypidiomorphic aggregates of olivine with enstatite, diopside, *etc.*[3] By some they have been regarded as very early intratelluric formations from the magma, by others as actual enclosed pieces of peridotites.

Some British examples. Several Tertiary examples of *tachylyte* have been described from Skye, Mull, and other Scottish islands, where the rocks occur as a local modification of more common types of basalt[4]. They usually enclose porphyritic crystals of olivine and magnetite, less commonly of augite and felspar. The glass is crowded with incipient growths of magnetite and occasionally of other minerals. These take the form of globulites, sometimes collected into cumulites (the Beal in Skye), of margarites (Lamlash near Arran), or of numerous minute opaque rods (Sorne in Mull, *etc.*), sometimes accompanied by transparent crystallites and belonites (Gribun in Mull). Spherulites occur in some instances. In the tachylyte of Ardtun in Mull[5] they are sometimes isolated, sometimes in bands, sometimes packed together, with polygonal boundaries, to the exclusion of any glassy matrix. When imperfect, they seem to consist of brown

[1] *Q. J. G. S.* (1886) xlii, 71, Pl. VII, fig. 3.

[2] Teall, *ibid.* (1884) xl, 235, Pl. XIII, fig. 1.

[3] For coloured figures see A. Becker, *Zeits. deutsch. geol. Ges.* (1881) xxxiii, Pl. III–V; Fouqué and Lévy, *Min. Micr.* Pl. XL, fig. 1.

[4] Judd and Cole, *Q. J. G. S.* (1883) xxxix, 444–462, Pl. XIII, XIV. For localities of numerous other examples in Mull, see Kendall, *G. M.* 1888, 555–560.

[5] Cole, *Q. J. G. S.* (1888) xliv, 300–307, Pl. XI.

globulitic matter, which is more condensed towards the centres. When better developed, they shew radiating fibres arranged in sectors, some brown and others grey, with pleochroism in both cases.

A different type of tachylyte has been described by Mr Groom[1] as occurring in narrow veins in association with the quartz-gabbro of Carrock Fell. A curious feature is the occurrence of clear spherical granules of quartz, as well as spherulitic aggregates of minute augite granules. These aggregates are generally surrounded by a narrow clear zone contrasting with the dark glass of the general matrix.

Of Recent tachylytes may be mentioned those of Mauna Loa in the Sandwich Islands. The only mineral common in phenocrysts is olivine. The matrix shews transitions from a pale yellow-brown glass with a few microlites only to one densely crowded with dark brown spherulites (often grown about an olivine or felspar crystal) and with opaque crystallites shewing beautiful skeleton outlines, star-like groupings, *etc.*[2]

Closely allied to the spherulitic tachylytes are the rocks known as *variolite*, of which examples have been described from Anglesey, the Lleyn district of Caernarvonshire, and various parts of Ireland[3]. The spherules shew considerable variety of structure, ranging from mere fan-like groupings of felspar microlites (*cf.* fig. 41, *A*) or sheaf-like aggregates with a lath-shaped crystal as nucleus (see Sollas) to very regular radiate spherulitic growths. They may be closely packed to make up the entire mass of a portion of the rock, or arranged in bands, or isolated in a matrix of brown or greenish glass with cumulites, globulites, *etc.* (see Cole). The individual spherules are commonly from one-tenth to one-

[1] *Q. J. G. S.* (1889) xlv, 298–304, Pl. XII.

[2] Dana, *Amer. Journ. Sci.* (1889) xxxvii, 450, 451, with figures.

[3] Miss Raisin (Lleyn), *Q. J. G. S.* (1893) xlix, 145–159, Pl. I; Cole (Careg Gwladys, Anglesey), *Sci. Proc. Roy. Dubl. Soc.* (1891) N.S. vii, 112–120, Pl. X; (Annalong, Co. Down) *ibid.* (1892) 511–519, Pl. XXI; (Dunmore Head, Co. Down) *ibid.* (1894) viii, 220–222; Sollas (Roundwood, Co. Wicklow) *ibid.* (1893) 99–106, figures. On foreign variolites see next note; also Löwinson-Lessing ('sordawalite') *M. M.* viii, 164 (*Abstr.*); Brauns (Hesse) *M. M.* ix, 255, 256 (*Abstr.*). For coloured figure of the 'variolite of the Durance' see Fouqué and Michel Lévy, Pl. XXIV, fig. 2.

half of an inch in diameter, but sometimes less or more. Secondary changes may cause devitrification of any glassy matrix, and give rise to a separation of iron-oxides, a production of epidote, *etc.* Variolite is found sometimes in small dykes, sometimes as a margin to larger basic intrusions or lava-flows, sometimes again in the interior of a diabase-mass, either bordering spheroidal joints or forming a selvage on irregular pillow-like portions into which the rock-mass is divided[1].

True basaltic lavas do not, as a rule, figure largely in the great volcanic groups which characterize the Lower Palæozoic in various parts of Britain. Sir A. Geikie[2] has noted olivine-basalts of early Cambrian (or late pre-Cambrian) age near St David's (Rhosson, Clegyr Foig, *etc.*). The idiomorphic crystals of olivine in these rocks are replaced largely by hæmatite. The ground-mass consists of augite-granules, abundant octahedra of magnetite, and a base crowded with globulites and trichites, felspar being only occasionally recognized. These characters suggest a resemblance to the limburgite type, noticed below.

Various basaltic lavas are intercalated in the Palæozoic strata of Cornwall and Devon. Some have been largely vitreous, the glass being now represented by a greenish yellow to brownish yellow serpentinous-looking substance which seems to be identical with the so-called palagonite (Cant Hill, near St Minver)[3]. This rock is often amygdaloidal.

Among the Ordovician lavas of the Lake District the Eycott Hill type is a very characteristic rock of wide distribution. It is a *hypersthene-basalt* or, as some petrologists would term it, on account of the absence of olivine, a basic hypersthene-andesite. It is quite distinct from the normal andesites of the same district, many of which also carry hypersthene. One variety, well developed at Eycott Hill,

[1] On this and other points see Cole and Gregory (M. Genèvre), *Q. J. G. S.* (1890) xlvi, 295–332, Pl. XIII; Gregory (Fichtelgebirge) *ibid.* (1891) xlvii, 45–62. On variolitic rocks near Fishguard, Pembrokeshire, see Reed, *Q. J. G. S.* (1895), li, with Pl. VI.

[2] *Q. J. G. S.* (1883) xxxix, 304, Pl. IX, fig. 4.

[3] Rutley, *Q. J. G. S.* (1886) xlii, Pl. XII.

has large rounded phenocrysts of labradorite with carlsbad and albite-twinning. These contain rather large opaque inclusions in the form of negative crystals and smaller enclosures with zonary disposition. In other varieties of the lavas these large crystals are not present. The ground-mass consists of slender striated prisms of plagioclase, crystals of hypersthene converted to pleochroic bastite, granules of augite, abundant

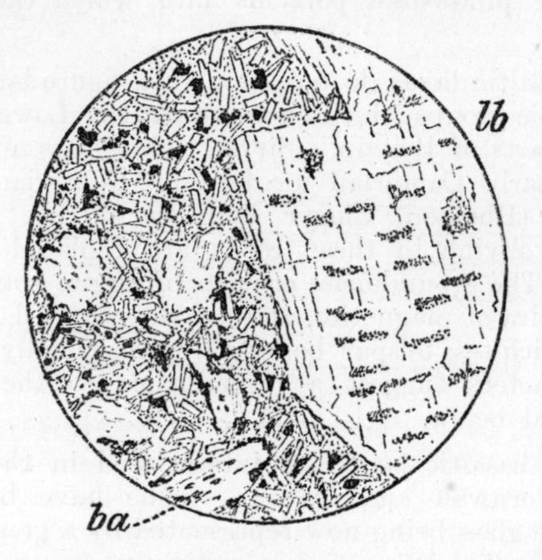

FIG. 43. HYPERSTHENE-BASALT, EYCOTT HILL GROUP, MELMERBY, CUMBERLAND; ×20.

To the right is one of the large crystals of labradorite (*lb*) with its peculiar inclusions. The hypersthene is represented by bastite pseudomorphs (*ba*): augite occurs in less abundance. These, with the little felspar-prisms, the granules of magnetite, and some residual glassy base, make up the bulk of the rock [1251].

magnetite, and an isotropic base (fig. 43). In the basic lavas of the Lake District generally olivine is entirely wanting. Hypersthene, pseudomorphed by bastite, is frequently present, but rarely to the exclusion of augite.

The basic lavas of the Carboniferous, on the other hand, are characteristically olivine-bearing rocks. Those of Derbyshire[1], locally known as 'toadstone,' are chiefly *olivine-*

[1] Arnold-Bemrose, *Q. J. G. S.* (1894) l, 611–625.

dolerites. The minerals present are idiomorphic olivine (sometimes replaced by a remarkable mica-like mineral[1]), augite, exceptionally a rhombic pyroxene (Sandy Dale), labradorite or a more basic felspar, magnetite and ilmenite. Most of the rocks are olivine-dolerites of granulitic structure, the augite occurring in grains (Castleton, Tideswell Dale[2], Miller's Dale, *etc.*). A few are ophitic (Peak Forest and Bonsall, see fig. 25, p. 115). Rarely there are porphyritic olivine-basalts with olivine and large augite phenocrysts in a ground-mass of small felspar laths, augite grains and prisms, and iron-ores, with little interstitial matter (Blackwell Lane, Great Low).

The Carboniferous basic lavas of the southern half of Scotland are generally *olivine-bearing basalts*, presenting a considerable variety of characters[3]. The commonest type has rather abundant small olivines and grains of augite in a mesh of slender felspars with microlitic augite and minute granules of magnetite (Dalmeny, Bathgate Hills, *etc.*). In another type the olivine phenocrysts are large, and the felspar microlites are found only in small amount (lowest lavas of Bathgate Hills, Linlithgowshire). A well-known rock from the Lion's Haunch on Arthur's Seat, Edinburgh[4], has numerous large, well-built crystals of augite, olivine, and felspar, with little crystals of magnetite, in a ground-mass of little crystals and microlites of felspar, granules of augite and magnetite, and some residual glass. In the lava of Craiglockhart Hill the ground-mass is more glassy, while the phenocrysts are augite and olivine without felspar. On the other hand, there is a holocrystalline type, which is an olivine-dolerite with granulitic to sub-ophitic structure (Gallaston, N.W. of Kirkcaldy). A curious variety, very rich in felspar, comes from Markle quarry in the Garlton Hills, Haddingtonshire[5]. Here olivine occurs only in small sporadic grains, while phenocrysts of labradorite are numerous, and the ground-mass consists of laths, microlites, and granules of felspar with dispersed magnetite and probably only a little augite.

[1] Arnold-Bemrose, *Q. J. G. S.* (1894) Pl. XXIV, figs. 1–4.
[2] Teall, Pl. IX.
[3] Geikie, *Q. J. G. S.* (1892) xlviii, Proc. 105, 106.
[4] Teall, Pl. XXIII, fig. 1.
[5] Hatch, *Trans. Roy. Soc. Edin.* (1892) xxxvii, 119, Pl. I, fig. 2.

A rock very like that of Lion's Haunch occurs as a dyke near the Stack of Scarlet in the South of the Isle of Man[1]. The phenocrysts are large idiomorphic crystals of fresh plagioclase and violet-brown augite, with pseudomorphs of calcite and serpentine after olivine. The ground-mass is of lath-shaped felspars, augite, and iron-ores. This is probably connected with the Carboniferous volcanic series of the Stack, which consists of tuffs with dykes and probably flows of a more compact basalt[2]. The latter is considerably decomposed, the augite being converted into chloritic and other products. Porphyritic felspars occur, and the little lath-shaped felspars of the ground-mass shew a fluxional arrangement. The much fresher basalt, which forms numerous small dykes in the south of the Isle of Man[3], is perhaps of Tertiary age. The olivine here is abundant and fresh, with inclusions of picotite. This, and sometimes plagioclase, are the only phenocrysts. The ground-mass is in general holocrystalline with fine texture, consisting of felspar microlites, ophitic violet-pink augite, magnetite, *etc.* Analcime, sensibly isotropic, occurs as a decomposition-product, as it does in the Salisbury Crags rock near Edinburgh and elsewhere[4].

In the neighbourhood of Limerick, in Ireland, is a considerable development of basaltic lavas of Carboniferous age. These differ from the Tertiary basalts of the same country in various points, and especially in frequently containing augite among the phenocrysts. Prof. Hull records hornblende in some examples.

The Tertiary basaltic rocks of the west and south of Scotland and the north-east of Ireland are olivine-basalts (including olivine-dolerites). They have been well described and illustrated by Prof. Judd[5], who has pointed out how the varied series of structures found in them constitute intermediate types between the holocrystalline plutonic rocks at the one extreme and the glassy basalts (tachylytes) at the

[1] Hobson, *Q. J. G. S.* (1891) xlvii, 443, 444.
[2] *Ibid.* 441.
[3] *Ibid.* 445–447.
[4] *Cf.* Teall, Pl. XXII, fig. 1.
[5] *Q. J. G. S.* (1886) xlii, 49–95, Pl. IV–VII: see also Teall, Pl. X.

other. He distinguishes two parallel lines of transition. One, characteristic of the true extruded lava-flows, includes the 'granulitic' dolerites and the basalts in which the augite tends to form granules between the felspar prisms ('micro-granulitic' structure). The other series of varieties includes the ophitic dolerites and the micro-ophitic basalts, in which the augite tends to enwrap and enclose the felspars: this seems to be the case especially in intrusive members of the group. The distinction is traced even through those basalts which consist largely of a glassy base, the crystallitic growths enclosed in the glass being in the one case in the form of granules and short microlites, often rounded, in the other case in the form of skeleton-crystals and more spreading growths.

Many of the Scottish dolerites and most of the basalts are porphyritic, the felspar occurring in two generations, of which the earlier is a thoroughly basic variety, near anorthite, while the later is less basic, approaching labradorite. Porphyritic augite, however, is not found, and this feature distinguishes the group of rocks in question from the Tertiary basalts of various European areas and also from many Carboniferous basalts of Scotland and Ireland. Professor Judd notes the absence of olivine-nodules as another distinctive feature of the British Tertiary basalts.

Of the Tertiary olivine-dolerites of intrusive occurrence in the Western Isles and others, probably of like age, in the Southern parts of Scotland, many have ophitic structures, and approach true diabases in their characters. Others, however, are of the 'granulitic' type, and these, in addition to the dominant lath-shaped felspars, shew a later generation of more acid composition in shapeless grains with marked zonary banding between crossed nicols (*e.g.* Craig Craggen in Mull, Muckraw in Linlithgowshire).

Volcanic rocks of ultrabasic composition seem to have a very limited development. To this place belong the *limburgites* of Rosenbusch (magma-basalts of Bořicky), lavas of highly basic nature, rich in olivine and augite and devoid of felspar. The only British examples yet recorded are from the Carboniferous of Scotland and Ireland. Dr Hatch[1] has described one from Whitelaw Hill near Haddington, which is in a very fresh condition. There are abundant well-shaped phenocrysts of olivine and augite, the latter having a very pale violet-brown tint in the interior, deepening towards the margin, with slight pleochroism. These minerals, with imperfect crystals of magnetite, occur in a ground-mass consisting of small augite-prisms set in brown to pale yellowish or colourless glass (fig. 41, *B*). Mr Watts has noted a limburgite in the Limerick district (Nicker), which closely resembles the preceding, though less perfectly preserved, the olivine being replaced by carbonates, *etc.* The augite has a strong zonary structure, the violet-brown tint being noticeable, while the interior of each crystal is paler or has a greenish colour. Augite in a second generation, magnetite granules, and more or less altered glass make up the ground-mass. From the same district Mr Hobson[2] has described the allied rock-type *augitite*, in which olivine as well as felspar is wanting. The rock consists essentially of two generations of augite and magnetite with some residual base, which has probably been glassy. These British examples are sufficiently like the typical rocks of Limburg in the Kaisertuhl, *etc.*, to render detailed description of these unnecessary. They are characteristically very basic lavas, in which crystallization has been arrested, both in the 'intratelluric' and in the 'effusive' period, before the separation of felspar had begun. The olivine is often a variety rich in iron, and becomes converted at the margin of the crystal into deep red hæmatite or brown limonite[3].

[1] *Trans. Roy. Soc. Edin.* (1892) xxxvii, 116–117, Pl. I, fig. 1.

[2] *G. M.* 1892, 348–350.

[3] Rosenbusch-Iddings, Pl. XIX, fig. 1: see also Fouqué and Lévy, *Min. Micr.* Pl. LII, fig. 2.

CHAPTER XV.

LEUCITE- AND NEPHELINE-BASALTS, ETC.

WE shall group together for convenience various basic and ultrabasic lavas in which leucite, nepheline, or, in certain types, melilite is a prominent constituent, with or without a lime-soda-felspar. In the phonolites and leucitophyres, described above, a potash-felspar was an essential mineral, and the rocks had other affinities with the trachytes. Although some of the rocks to be noticed resemble the phonolites and leucitophyres in some features, they are for the most part allied rather with the basalts, while the varieties having any considerable amount of glassy base graduate into the limburgites and augitites.

The rocks in which leucite or nepheline only partly takes the place of felspar are termed *leucite-* or *nepheline-tephrites* when free from olivine, and *leucite-* or *nepheline-basanites* when containing that mineral. For those rocks which have the felspathoid mineral to the exclusion of felspar the name *leucitite* or *nephelinite* is used when olivine is absent, and *leucite-* or *nepheline-basalt* when olivine is present. In all these divisions the leucite-bearing and the nepheline-bearing types are on the whole distinct, though the rocks characterized by either of the minerals may contain the other as an accessory.

To these types may be added the *melilite-basalts*, in which the mineral named is abundant, usually with little or no felspar and with abundant olivine. Rosenbusch separates from the lavas, under the name alnöite, a rock which occurs in dykes in association with nepheline-syenite.

The rocks here noticed are scarcely known among the older lavas. A Palæozoic leucitite has been described from Siberia[1], and a Palæozoic melilite-basalt from Canada.

Constituent minerals. The *leucite* of these rocks may be in two generations, differing in size. The crystals are always idiomorphic icositetrahedra, but often more or less rounded. They usually shew feeble birefringence and the characteristic lamellar twinning. Augite microlites and granules, glass-inclusions, *etc.*, are often arranged in zones, or grouped in the centre of the crystal[2].

The *nepheline* in the porphyritic types is usually confined to the ground-mass. In the nephelinites and nepheline-basalts it is commonly idiomorphic, except in some of the holocrystalline rocks. In other cases it often forms small allotriomorphic crystals, not easily identified, and its distribution may be local. Its alteration-products are natrolite and other zeolites in radiating aggregates.

Other felspathoid minerals, *sodalite*, *haüyne*, and *nosean* are not uncommon as phenocrysts in the rock-types richest in leucite and nepheline, but they occur only as accessories.

The yellow or colourless *melilite*[3] is recognised by its weak double refraction, straight extinction, and peculiar micro-structure. Idiomorphic crystals have a tabular habit parallel to the base, and the basal faces sometimes form concave-curves. The mineral may also be quite allotriomorphic, and, when it occurs as an accessory in leucite-lavas, has sometimes the form of a framework enclosing other minerals in pœcilitic fashion (fig. 45).

[1] Chrustchoff, *M. M.* x, 177 (*Abstr.*).
[2] See Rosenbusch-Iddings, Pl. VIII, fig. 2; XIV, fig. 5; Zirkel, *Fortieth Parallel*, Pl. I, figs. 21–23, V, fig. 4.
[3] See Rosenbusch-Iddings, pp. 159, 160; Pl. XIV, fig. 6; XV, fig. 6; Adams, *Amer. Journ. Sci.* (1892) xliii, 277, 278; Smyth, *ibid.* (1893) xlvi, 104–107; Osann, *Journ. of Geol.* (1893) i, 342, 343.

This latter mode of occurrence is sometimes seen also in the *sanidine* which occurs as an accessory in some of the leucite- and nepheline-lavas, linking them with the leucitophyres and phonolites. The *plagioclase* felspars, which are found in some types of these rocks, are always of a basic variety. There may be phenocrysts with idiomorphic outline, tabular habit, albite-lamellation, zonary structure, and zones of glass-inclusions; while the felspars of the ground-mass vary from narrow laths, often only once twinned, to mere microlites. These shew a tendency to spherulitic arrangement, and the phenocrysts too may form radially grouped aggregates (fig. 44).

The usual coloured constituent in the rocks here considered is *augite*. It often occurs in two generations, the earlier relatively large and well shaped. The colour is commonly green, but often varies in concentric zones, becoming sometimes pale violet, with distinct pleochroism, at the margin of a crystal. Again, there are sometimes two kinds of porphyritic augite differently coloured. Some nephelinites have a purple-brown pleochroic 'hour-glass' augite. Exceptionally some of the rocks contain little yellowish green needles of *ægirine*. A brown or red-brown or red *biotite* is very common in the nepheline- and melilite-rocks, often shewing resorption-phenomena. Brown *hornblende* is an occasional accessory in some rocks, and commonly shews a corrosion-border of magnetite and augite[1].

Olivine is an essential constituent in many of the types, and has the same general characters as in basalts. In some of the most basic rocks the mineral is a hyalosiderite, and often becomes red by the separation of iron-oxide.

Iron-ores are commonly present, and in the olivine-bearing rocks often abundant. They are *magnetite* and *ilmenite*, the latter sometimes in deep brown translucent scales.

Apatite is a pretty constant accessory. In some of the nepheline-dolerites it builds large crystals. A pale violet or blue tint, with evident dichroism, is sometimes seen. Some

[1] Rosenbusch-Iddings, Pl. XIV, fig. 1.

of the leucite- and nepheline lavas have *melanite*-garnet, brown in slices and always isotropic. A very common accessory in the melilite-basalts and some nepheline-rocks is *perofskite* in minute octahedra, shewing in high relief in consequence of their refractive index[1].

Leading types. Our illustrations must be drawn entirely from foreign sources, since, with the exception of the few phonolites already noted, no lavas containing felspathoid minerals are found within the British area.

It must be noticed that the several types to be distinguished are not always sharply marked off from one another. This is especially the case with the felspar-bearing members, the tephrites and the basanites having in great measure the same general characteristics, except for the not very considerable proportion of olivine in the latter. The differences between the leucitites and nephelinites on the one hand and the leucite- and nepheline-basalts on the other are, however, more marked, the olivine-bearing types being notably richer in the ferro-magnesian constituent (augite) and in iron-ores. Among rocks characterized specially by melilite, the only important type is melilite-basalt, containing abundant olivine and typically no felspar.

A well-known *leucite-tephrite* comes from Tavolato near Rome. It is remarkable for an abundance of blue haüyne. There are two generations of leucite, both shewing twin-lamellation. A greenish brown ægirine occurs as well as augite. Both lath-shaped plagioclase and sanidine are found, the latter sometimes occurring as an interstitial matrix to the other minerals, though in other examples there is some glassy residue. The rock also contains grains of melanite. Other leucite-tephrites occur in the Kaisertuhl (near Freiburg in the Breisgau), in Bohemia, *etc.* The Bohemian examples contain no haüyne, and have leucite for the most part confined to the holocrystalline ground-mass.

The lavas of Vesuvius[2] stand between leucite-tephrite and

[1] Rosenbusch-Iddings, Pl. xv, fig. 2.

[2] See Fouqué and Lévy, *Min. Micr.*, Pl. XLIX, fig. 1; Haughton and Hull, *Trans. Roy. Ir. Acad.*, xxvi, Pl. II.

leucite-basanite, olivine being, as a rule, not very abundant. The conspicuous phenocrysts are of leucite (with inclusions of brown glass and augite-microlites), plagioclase (often in radiating groups of crystals), augite, and usually olivine ; and the same minerals, except the last, recur as constituents of the ground-mass. Magnetite and apatite are always present, and in some cases biotite is plentiful. Nepheline, sanidine, and brown hornblende are rarer, and sodalite is confined to crevices, where it seems to have been formed after the consolidation of the rock. The ground-mass is usually holocrystalline or with only a little brownish or yellowish glass, but there are vitreous[1] and pumiceous modifications. The lavas of different eruptions, while differing in the relative proportions of some constituents, preserve the same general type[2] (fig. 44).

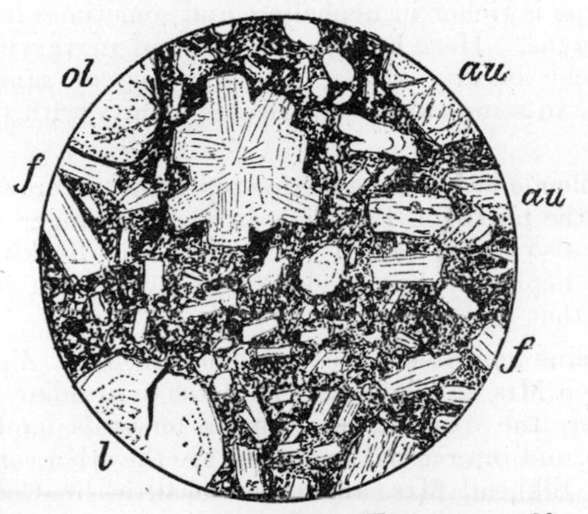

FIG. 44. LEUCITE-BASANITE, VESUVIUS; ×20.

This shews leucite (*l*) and crystals or groups of felspar (*f*), both with zones of inclusions, augite (*au*), olivine (*ol*), magnetite, and a little isotropic residue [845].

The rock described by Hague[3] from the Absaroka range in Wyoming is an example of a leucite-basanite having affinities

[1] Fouqué and Lévy, *Min. Micr.*, Pl. LI, fig. 2.
[2] See, *e.g.*, Matteucci on lavas of 1891, *M. M.*, x, 181, 182 (*Abstr.*).
[3] *Amer. Journ. Sci.* (1889) xxxviii, 45.

with the leucitophyres. Olivine and augite are porphyritic in a ground-mass essentially of leucite and sanidine, plagioclase being only scantily represented. Magnetite, apatite, and a little mica are present, and there may be a very small proportion of glassy base.

The scoriaceous lava of Niedermendig, in the Laacher See district, is placed by Rosenbusch between the leucite- and the nepheline-tephrites. He considers the conspicuous blue haüyne crystals to be of foreign derivation.

The lavas of the Canary Islands afford a great variety of *nepheline-tephrites* and nepheline-basanites, the former predominating. Some of these are designated the 'basaltoid' type, and in these nepheline is not present in any large proportion. The structure is usually holocrystalline. The 'phonolitoid' type is richer in nepheline, and sometimes has blue or yellow haüyne. Here hornblende is found in varying proportion, sphene occurs, and a predominance of sanidine over plagioclase in some varieties indicates affinity with the phonolites.

Hornblende-bearing nepheline-tephrites occur also in the Rhön (to the north of Bavaria), in the Thüringer Wald, *etc.* There are also rocks, named 'basanitoid' by Bücking, having no actual nepheline but a glassy base very rich in soda to represent that mineral.

Nepheline-tephrites have been described by Zirkel[1] from the Kawsoh Mts. in Nevada. These have sanidine predominating over the plagioclase: augite crystals and needles, magnetite, and interstitial nepheline are the other constituents. From the Elkhead Mts. and other localities in Colorado the same writer[2] notes examples of *nepheline-basanite.* One type, of coarse texture, has large crystals of olivine, idiomorphic zoned augite, plagioclase, and interstitial nepheline. Magnetite is plentiful, and biotite is often present. A nepheline-basanite from Southern Texas[3], on the other hand, is of the type poor in olivine, carrying brown hornblende among the phenocrysts and sanidine in the ground-mass.

[1] *Micr. Petr. Fortieth Parallel* (1876), 255, 256.
[2] *Ibid.* 256–258.
[3] Osann, *Journ. of Geol.* (1893) i, 344–346.

Nepheline-basanites in considerable variety are associated with the nepheline-tephrites of the Rhön, the Canaries, *etc.* Some are poor in nepheline and felspar, and approximate to the limburgites. Doelter's 'pyroxenite' (augitite) from the Cape Verde Islands is similar, having only crystals of augite and some magnetite in a glassy ground-mass of composition agreeing with nepheline.

Good examples of the type *leucitite* come from the Alban Hills near Rome (Capo di Bove[1], *etc.*). They are non-porphyritic rocks, very rich in leucite and relatively poor in augite. Other constituents are brown biotite, yellow striated melilite, and clear sanidine, all of which occur in crystal-plates enclosing the leucite and augite in pœcilitic or ophitic fashion (fig. 45).

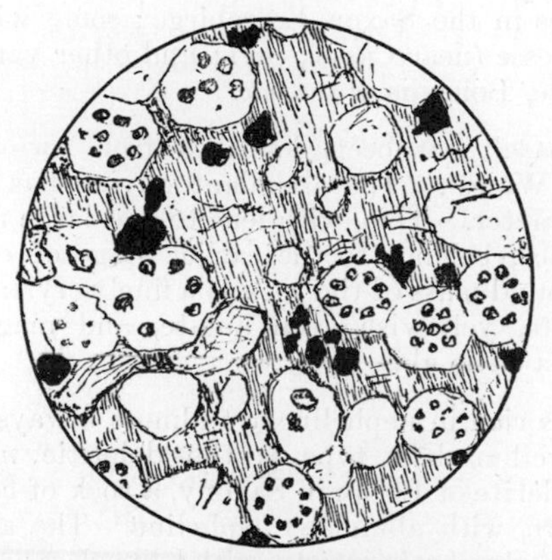

FIG. 45. LEUCITITE, CAPO DI BOVE, NEAR ROME; ×100.

Small leucites with zonally grouped inclusions are numerous, and augite and magnetite also occur. All these are enclosed by a large crystal of yellowish striated melilite. In other parts of the slide sanidine plays a similar part [G 243].

The rock described by Zirkel[2] from the Leucite Hills,

[1] See Fouqué and Lévy, *Min. Micr.*, Pl. L, fig. 1; Rosenbusch-Iddings, Pl. XIV, fig. 6. A type richer in augite, from Frascati, is shewn in Pl. L, fig. 2, of *Min. Micr.*

[2] *Micr. Petr. Fortieth Parallel*, 260, 261; Pl. V, fig. 4; I, figs. 21–23.

Wyoming, is even richer in leucite. In addition to this mineral, it contains only a pale biotite, scattered needles of green augite, apatite, and a small quantity of magnetite.

Leucitites very rich in blue haüyne occur in the Cape Verde Islands; varieties rich in augite in the Kaisertuhl, in Brazil, *etc.*

Of *leucite-basalt* good examples come from the Eifel district (Fornicher Kopf, Hummerich, *etc.*). These have phenocrysts of olivine, augite, and often biotite, in a ground-mass which is always very fine-grained but rarely contains any glassy residue. It consists of predominating augite with leucite and often nepheline, while a little sanidine sometimes occurs interstitially. Leucite-basalts, often with some melilite and biotite, occur at various places in the Saxon Erzgebirge; some with accessory haüyne in Hesse (near Cassel, *etc.*); and other varieties in the Vogelsgebirge, Bohemia, Java, *etc.*

Leucite-basalt has been described from two localities in New South Wales[1]. The abundant olivine has a somewhat peculiar character. This, with leucite and large ragged flakes of yellow mica, belongs to the earlier stage of consolidation, while the ground-mass of the rock is a finely-crystalline aggregate of leucite, yellowish-green augite, and magnetite, with occasionally a little glass.

The rocks rich in nepheline are almost always holocrystalline. A well-marked type is the doleritic *nephelinite* or nepheline-dolerite of Löbau in Saxony, a rock of comparatively coarse texture, with abundant nepheline. The augite is of a purple-brown pleochroic variety, with hour-glass or other zonary growth, and often idiomorphic (fig. 46). Locally the structure of the rock may become intersertal or, again, micrographic. Besides the abundant nepheline, subordinate sanidine may occur, and more rarely a plagioclase. The common iron-ore is a titaniferous magnetite, and apatite needles occur abundantly. In the otherwise similar type of Meiches, in the Vogelsberg (Hesse), leucite, in irregular grains crowded with

[1] Judd, *M. M.* vii, 194, 195 (1887); Edgeworth David and Anderson, *Rec. Geol. Surv. N. S. W.* (1890) i, 159–162, Pl. XXVIII.

apatite needles, becomes a prominent constituent. Both rocks shew transitions to nepheline-basalt, of finer texture, with less

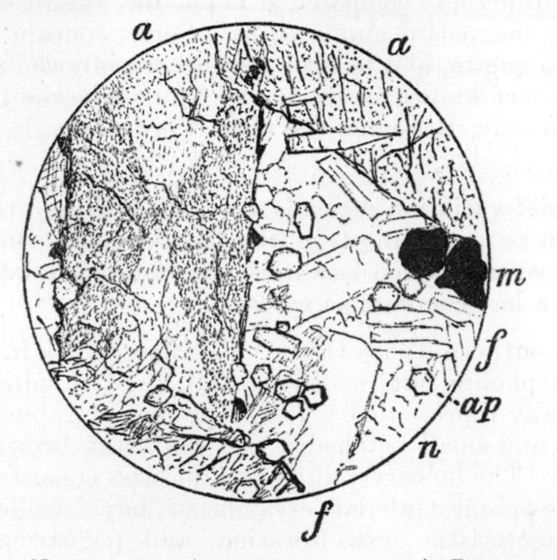

FIG. 46. NEPHELINITE (NEPHELINE-DOLERITE), LÖBAUER BERG, SAXONY; × 20.

The minerals shewn are nepheline (*n*), some felspar (*f*), purplish-brown augite (*au*) with hour-glass structure, magnetite (*m*), and apatite (*ap*), the rock being holocrystalline. The coming in of felspar marks a transition to the tephrite type [G 220].

nepheline and with abundant phenocrysts of olivine. The same is true of another well-known nephelinite, that of Katzenbuckel in the Odenwald (Baden).

Another type ('basaltic nephelinite') occurs in the Grand Canary, *etc.*, and by the coming in of plagioclase passes into the tephrites. It is of fine texture and much richer in augite than the preceding. Varieties, some rich in haüyne, occur in the Erzgebirge and in Bohemia. Rosenbusch's 'phonolitic' type is, on the other hand, poor in coloured minerals, and carries no augite-phenocrysts. In the frequent presence of ægirine-microlites, the abundance of idiomorphic nepheline, and the coming in of sanidine, this type approaches the phonolites.

The *nepheline-basalts*, much more widely distributed than nephelinites, shew less variety of character. They are typically holocrystalline rocks composed of nepheline, augite, and olivine, with some magnetite and apatite. Some contain biotite in addition to augite, and haüyne may accompany the nepheline[1]. Such rocks are known in Hesse and Thuringia, the Eifel, many parts of Saxony, Bohemia, the Cape Verde Islands, and other regions.

The chief variation depends upon the coming in of melilite in addition to nepheline (*e.g.* Herchenberg and Bongsberg in the Eifel, several Saxon localities, the Sandwich Islands, *etc.*). Leucite is a less common accessory.

From southern Texas Osann[2] describes a rock intermediate between nepheline- and melilite-basalt, the two minerals being about equally represented. Large olivines are abundant, with magnetite and small octahedra and grains of brownish violet perofskite. The holocrystalline ground-mass consists of abundant augite-prisms, tabular crystals of faint yellow melilite with characteristic cross-fibration and peg-structure (Ger. Pflockstructur), and aggregates of shapeless grains of nepheline. Felspar is entirely wanting.

In the true *melilite-basalts* nepheline is wanting or at most an accessory. Phenocrysts of olivine, augite and biotite are imbedded in a usually holocrystalline ground-mass of smaller biotite, zoned augite, sometimes olivine, and melilite. The last sometimes occurs also among the phenocrysts. Biotite is specially characteristic, and in the first generation may form quite large flakes. Rocks answering to this description are known from Hochbohl, near Owen, and Urach in Wurtemburg, from Görlitz in the Prussian Lausitz, as dykes on Alnö, an island off the coast of Sweden, *etc.* A good example, of Silurian age, is described from Ste Anne de Bellevue near Montreal[3]. Here the phenocrysts are biotite in large and abundant crystals, olivine more or less converted to hæmatite, and augite: the ground-mass is of biotite, olivine, augite, magnetite, and meli-

[1] Fouqué and Michel Lévy, Pl. XLIX, fig. 2.
[2] *Journ. of Geol.* (1893) i, 341–343.
[3] Adams, *Amer. Journ. Sci.* (1892) xliii, 269–279.

lite, with apatite and perofskite, the last a mineral rarely absent from such rocks. The melilite is the latest product of consolidation, forming imperfect crystals of tabular habit with the characteristic 'peg-structure.' A rock from Mannheim, N.Y.[1], differs from this chiefly in the absence of pyroxene, and both closely resemble the typical 'alnöite' of Alnö, off the coast of Sweden, which also contains augite in addition to the large phenocrysts of brown mica.

[1] Smyth, *Amer. Journ. Sci.* (1893) xlvi, 104–107.

D. SEDIMENTARY ROCKS.

UNDER the head of sedimentary rocks we shall include the stratified deposits formed for the most part, though not exclusively, under water by the accumulation of detritus and of fragmental material of volcanic origin, by organic agency, and by chemical action or the evaporation of saline solutions. The last clause includes the secondary cementing material of many fragmental rocks, as well as the less common deposits of rock-salt, *etc.*, which do not demand special notice.

The rocks exhibit great variety of composition and characters, and in the nature of the case do not admit of any very strict petrological classification. They will be treated mainly under four groups: the coarser detrital deposits (*arenaceous*), the finer detrital deposits (*argillaceous*), the rocks consisting essentially of carbonate of lime (*calcareous*), and the fragmental volcanic rocks (*pyroclastic* of some authors). In all, with the exception of some of the calcareous rocks, a fragmental or '*clastic*' structure is essentially present: this, with the bedded occurrence, may be taken as characteristic of the whole.

CHAPTER XVI.

ARENACEOUS ROCKS.

The arenaceous rocks are typical fragmental ('clastic') accumulations, consisting of grains of one or more materials mechanically derived, to which may be added interstitial matter deposited in place. There is thus a distinction between original or 'allothigenous' constituents, derived from a distance, and secondary or 'authigenous' constituents formed after the accumulation of the grains. The fragmental nature of the rocks is usually evident to the eye, and the conditions of deposition in water may be indicated by an appearance of lamination, but this is rarely so well marked as in some argillaceous rocks.

The name *sand* (Fr. sable) is reserved for incoherent deposits: when compacted by some cementing medium, they become *sandstone* or *grit*. These last two words are often used synonymously, though different writers have employed them to mark various distinctions. If a distinction be made, it is perhaps best to name the round-grained rocks sandstones, and those with angular grains grits. Such epithets as felspathic and calcareous are used to describe the nature sometimes of the grains, sometimes of the cement: they usually need no explanation. The old term *greywacke* (Ger. Grauwacke) has been revived for a complex rock with grains of quartz, felspar, and other minerals and rocks united by a cement usually siliceous. An *arkose* is a deposit derived from the direct destruction of granite or gneiss, and containing abundant felspar. A *quartzite* (of the type belonging here) is a rock consisting of grains chiefly of quartz with a quartz cement.

The coarsest clastic deposits, in which pebbles occur as well as grains, are named *conglomerate* or pudding-stone (Fr. poudingue) when the large fragments are rounded, and *breccia* (Fr. brèche) when they are angular. These rocks will require but little notice.

Derived grains[1]. Since most sands are derived directly or indirectly (*i.e.* through the medium of earlier sedimentary deposits) from the waste of igneous or crystalline rocks, the *most usual minerals* in sand-grains are those which figure largely in the composition of large bodies of rock, such as granites, gneisses, and crystalline schists. But chemical processes tend to make a selection among these constituents, for the material is commonly affected by partial decomposition, either prior to the disintegration of the parent rock-masses, during transport, or subsequently to the accumulation of the clastic deposit. So the commonest constituents of sands are those abundant rock-forming minerals which are least prone to chemical changes, such as quartz and white mica. Felspars, augite, hornblende, and dark micas may occur plentifully in particular deposits, but are less characteristic of sands in general, while unstable minerals like olivine rarely occur among detrital material. Certain accessories, such as zircon and rutile, are widely distributed in sands, but only in small quantity. Others may be abundant locally, just as the modern sands on our coasts are found in particular localities to be rich in garnet, or flint, or tourmaline, or ilmenite (menaccanite)[2]. The admixture of few or many constituents depends on the extent and geological diversity of the drainage-area from which the material was derived. River- and lake-sands usually shew less variety than those of marine origin[3].

Some coarse-grained deposits contain composite *rock-fragments*, *e.g.*, a piece consisting of quartz and felspar with the

[1] For much information on sand-grains see Sorby, *Presid. Address*, *Q. J. G. S.* (1880) xxxvi, *Proc.* pp. 47–65; also *Anniv. Address Micro. Soc.* (1877) *Monthly Micro. Journ.*

[2] The heavier accessories may be separated from loose sands by levigation, *etc.* For a dry method, see Carus-Wilson, *Nature* (1889) xxxix, 591.

[3] See Julien and Bolton, *Proc. Amer. Assoc.* (1884) 413–416.

relations characteristic of granite. Other sandstones have numerous fragments of lava. Recent deposits near the volcanic islands of the Pacific sometimes consist wholly of rolled fragments of lava, pieces of decomposing volcanic glass (palagonite), small chips of pumice, *etc.* By admixture of material of *directly* volcanic origin these volcanic sands graduate into tuffs.

The accumulations composed mainly or entirely of organic fragments (shell-sands, coral-sands, *etc.*) are more conveniently placed with the limestones.

The *form and superficial characters* of sand-grains, best studied by mounting the material dry or in water, may depend upon the properties of the individual minerals and their mode of occurrence in the parent-rocks; upon the effects of attrition during transport; and sometimes upon crystalline growth subsequent to the accumulation of the deposit. Grains of felspar, hornblende, *etc.*, usually have their boundaries partly determined by the cleavages of the mineral; mica tends to form flat flakes or scales; minerals like zircon and anatase, which in the parent-rock built small well-formed individuals, often preserve their form intact. They are probably released in some cases by the destruction in the sand itself of an enclosing mineral, such as biotite. Quartz breaks into fragments of irregularly angular outline. If originally of interstitial occurrence (*e.g.* in a granite) it partly retains its highly irregular contour, and the minor irregularities produce a rather opaque appearance on the surface. Quartz-grains from a fine mica-schist, on the other hand, tend to flaky and lenticular shapes.

The degree of *rounding* produced by attrition during transport depends on the hardness of the mineral, but also on the nature and duration of the mechanical agencies involved. Large grains are often more rounded than small (fig. 47). Marine sands are in general more round-grained than those of rivers and lakes, while wind-borne sands, such as those of deserts, are still more rounded by friction. Only in these last are the smallest grains ever found to be well rounded.

It is usually possible to form some opinion as to the source or *sources of the derived material* of a sand. The minerals identified give a clue to the parent rock or rocks, and special features in the minerals may also afford information. Thus the existence of fluid-, glass-, or other cavities in crystal-fragments, the presence of rutile-needles in quartz-grains, *etc.*, may tell us whether the minerals in question originally formed part of a plutonic, a volcanic, or a metamorphic rock, or of several different rocks. Too much stress must not be laid on the rounding of grains as indicating the distance of their source. Long-continued drifting to and fro within a limited area may cause more attrition than many thousand miles of travel in one direction: further, friction is much more effective under subaërial than under subaqueous conditions. Again, sand-grains may be furnished by the destruction of older arenaceous deposits.

The *coarseness or fineness* of sandstones may vary considerably. The sifting action of running water tends to collect in one place grains of roughly equal dimensions, but some sandstones contain grains of two very different sizes, the smaller occupying the interspaces between the larger (fig. 47). A very

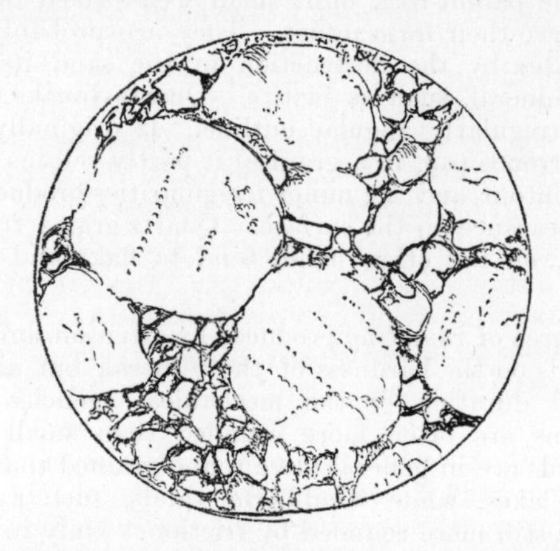

FIG. 47. 'TOP GRIT' OR UPPERMOST BED OF THE QUARTZITE SERIES, NEAR INCHNADAMFF, SUTHERLAND; ×20: shewing small angular quartz-grains occupying the interstices between the larger rounded ones [1665].

common size for the grains of quartz and felspar in many sandstones is from ·01 to ·03 inch[1].

Authigenous constituents. In addition to the clastic grains, sandstones and grits contain material deposited upon the surfaces of the grains, or filling in partially or wholly the interstices between them, and thus serving to bind them into a coherent rock. Whether formed by the recrystallization of calcareous or other matter laid down with the detritus, by the redeposition of material dissolved from the grains themselves, or by the introduction in solution of some extraneous substance, this cement must be regarded as formed in place, and its accumulation constitutes a new chapter in the history of the rock. The cementing medium itself is usually calcareous, ferruginous, siliceous, or some mixture of these.

The *calcareous cement* has probably been in most cases deposited in the form of mud, comminuted shells, *etc.*, with the original grains, but it becomes effective as a binding material only after some amount of solution and redeposition, which commonly gives it a more or less evident crystalline texture. Exceptionally a crystalline growth of calcite may enclose grains in ophitic or pœcilitic fashion, as in the Fontainebleau Sandstone of the Paris Miocene, but usually the calcareous cement is strictly interstitial, and it does not always fully occupy the interspaces between the grains. In rare cases other salts, such as gypsum, may serve as a cement.

Many sandstones are cemented by *ferruginous* matter or a mixture of ferruginous and calcareous. The red oxide and the brown hydrated oxide of iron occur in this way. Frequently the oxide forms a thin coating or pellicle round each grain of sand. The pellicle can be removed by acid, leaving the grains colourless.

The clayey material (kaolin, very fine mica, *etc.*), which occurs interstitially in some sandstones, is probably to a great extent authigenous, representing the decomposition of felspar grains, *etc.* Similarly a chloritic mineral is not uncommon,

[1] See Bonney, *Rep. Brit. Assoc.* for 1886, p. 601, and *Nature* (1886) xxxiv, 442.

and may be derived from the destruction in place of such minerals as hornblende and biotite.

In the tougher sandstones and grits the cementing matter is in the main *siliceous.* When the grains are angular and of various sizes, the interspaces may be very small, and the interstitial silica, concealed by the grains and perhaps by kaolin dust or iron-staining, may be difficult to observe. In more or less porous rocks, the little cementing matter required may be provided by some slight solution of the quartz-grains themselves at the points where they press on one another, as suggested by Mr Wethered for the sandstones of the Bristol coalfield.

When spaces have existed between the original grains, it is usually seen that the siliceous cement has been deposited in crystalline continuity with the original quartz as a *new outgrowth of the clastic grains.* This secondary enlargement of the grains is verified by the new material extinguishing simultaneously with the old between crossed nicols. Again, many sandstones which have not been compacted into hard rocks exhibit a similar new growth on the surfaces of the grains, and in this case (fig. 48) the added material often shews good

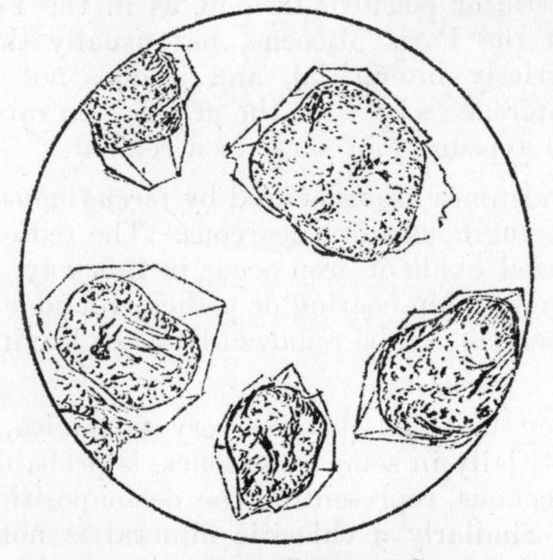

FIG. 48. QUARTZ-GRAINS FROM PENRITH SANDSTONE, PENRITH BEACON; ×20: shewing a secondary outgrowth of quartz with crystal-faces [1920].

crystal faces[1] ('crystallized sand'). The enlargement is commonly clearer than the nucleus, and the division between them is marked by a line of dusty inclusions or by a thin partial coating of some deposit older than the outgrowth. Though characteristic of quartz, a similar outgrowth is occasionally found on fragments of felspar[2] and hornblende[3].

In less frequent examples new-formed quartz has a radial arrangement about original grains, or is oriented independently. Again, a cement of cryptocrystalline or chalcedonic silica is known in some rocks. This, however, is rather characteristic of volcanic sandstones and conglomerates in regions of hot-spring action: *e.g.* in the Yellowstone Park rolled fragments of obsidian and rhyolite are thus cemented into a hard rock.

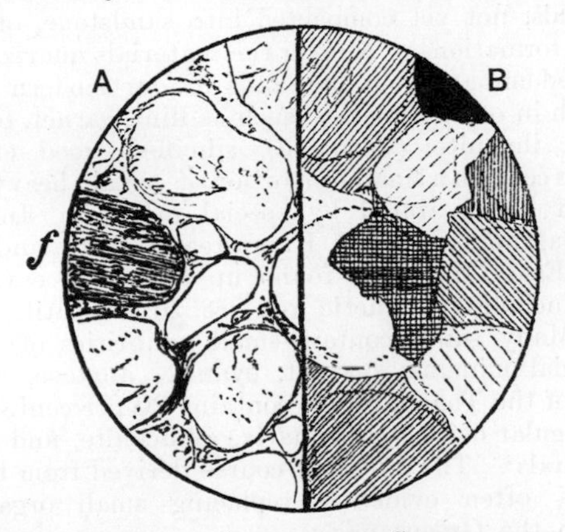

FIG. 49. QUARTZITE, STIPERSTONES, SHROPSHIRE; ×50: *A* in natural light, *B* between crossed nicols. The grains are of rolled quartz with an occasional turbid felspar (*f*), and the interspaces are filled by a secondary outgrowth of quartz from the grains. The shading is diagrammatic, to indicate different interference-tints. A composite grain in the centre shews outgrowths from both portions [224].

[1] Sorby (*Address cit. supra*, 62–64). For figures see R. D. Irving, 5th *Rep. U. S. Geol. Surv.* (1885) Pl. xxx.

[2] Irving, *ibid.* pp. 237–241.

[3] Van Hise, *Amer. Journ. Sci.* (1885) xxx, 232–235.

When a deposit originally a quartz-sand becomes completely compacted by an interstitial cement of secondary quartz, the result is a *quartzite* of the ordinary type. Such rocks often consist wholly or almost wholly of quartz, but in a thin slice the distinction between the derived grains and the interstitial cement comes out clearly. Usually the new quartz is a crystalline outgrowth from the grains, the space between two grains being occupied by quartz of which part is in continuity with one grain, part with the other. Between crossed nicols the slice therefore assumes the appearance of an irregular mosaic[1] (fig. 49).

Some British examples[2]. The forms and general characters of sand-grains may be studied in modern deposits and in the sands, not yet compacted into sandstone, of the later geological formations. Among the materials quartz, as a rule, largely predominates, but the sands of our modern coasts are locally rich in other minerals, such as flint, garnet, tourmaline, magnetite, ilmenite (Cornwall), silicified wood (Eigg), *etc.* Most sands contain a small proportion of certain heavy minerals, which can be separated by special methods. In the fine-grained Bagshot Sands of Hampstead Heath and of High Beech in Essex Mr Dick[3] found up to 4 per cent. of dense minerals, including magnetic iron ore, zircon, rutile, and tourmaline. Many sands contain small quantities of these and other special minerals (garnet, cyanite, anatase, *etc.*). The basal bed of the Thanet Sands contains 20 per cent. of flint in sharply angular chips, with quartz, glauconite, and numerous other minerals[4]. The flint is of course derived from the Chalk. Glauconite, often evidently replacing small organisms, is common in the Greensands.

[1] For coloured figures see Teall, Pl. XLV, fig. 2; XLVI, fig. 1; Irving (*cit. supra*), Pl. XXXI; Irving and Van Hise, *On Secondary Enlargements of Mineral Fragments* (1884), *Bull. No.* 8 *U. S. Geol. Surv.*, Pl. III–VI.

[2] Interesting information concerning British arenaceous rocks is contained in Sorby's *Presidential Address*, quoted above, and earlier papers (*Proc. Yorks. Geol. and Polyt. Soc., etc.*). See also J. A. Phillips, *Q. J. G. S.* (1881) xxxvii, 6-27; Bonney, *Nature* (1886) xxxiv, 442–451, and *Rep. Brit. Assoc.* for 1886, 601-621.

[3] *Nature* (1887) xxxvi, 91, 92; Teall, Pl. XLIV; *cf.* Fouqué and Lévy, Pl. IV.

[4] Miss Gardiner, *Q. J. G. S.* (1888) xliv, 755–760.

The form of quartz-grains depends in great measure upon their source, whether directly from crystalline rocks or from older sandstones or grits. Thus the Glacial sands of the Yorkshire coast, which must come chiefly from crystalline rocks, have sharply angular shapes, and the grains on the modern beaches of that coast, most of which are doubtless washed out of the Glacial accumulations, are scarcely more abraded. On the other hand, modern sands on the south-east coast of England, derived very largely from older arenaceous deposits, have a considerable proportion of rounded grains. On the north-west coast both Glacial and modern sands often contain extremely rounded grains, explained as being derived from the 'millet-seed' sandstones of the Trias, but these are mixed with angular quartz in various proportions. The grains of the sand-dunes on our coasts are much less rounded than those of desert sands.

The Mesozoic formations afford numerous examples of calcareous and ferruginous cements. Thus the Calcareous Grits of Yorkshire have a cement of calcite, often stained or mixed with iron-oxide, and some of them might with equal propriety be named impure gritty limestones. The Kellaways Rock has usually a ferruginous cement. Specimens shew angular to subangular grains of quartz about ·02 inch in diameter set in an opaque brown framework (fig. 50, *B*). In the Lower Greensand of the eastern counties the cement is sometimes largely ferruginous or with a little interstitial quartz ('carstone' at Hunstanton), but in many cases is of granular calcite, which may be iron-stained. Occasionally the calcite builds large plates enclosing many of the partly rolled quartz-grains, *etc.*, as in the Fontainebleau Sandstone, but without crystal-faces[1] (Spilsby in Lincolnshire, Copt Point near Folkestone). Many of these rocks have little grains of bright green glauconite with various rounded shapes, explained as casts of foraminifera. Another feature is the occurrence of little round oolitic grains of dark brown iron-ore ('carstone' of Hunstanton, and Roslyn Hill, Ely). These grains have a concentric shell structure, and, when dissolved in acid, leave a siliceous skeleton

[1] Mr Watts notes the same feature in the Kellaways Rock of Bedfordshire, *Rep. Geol. Assoc.* (1894) xiii, 360.

(fig. 50, *A*). Zircon crystals are among the denser constituents[1].

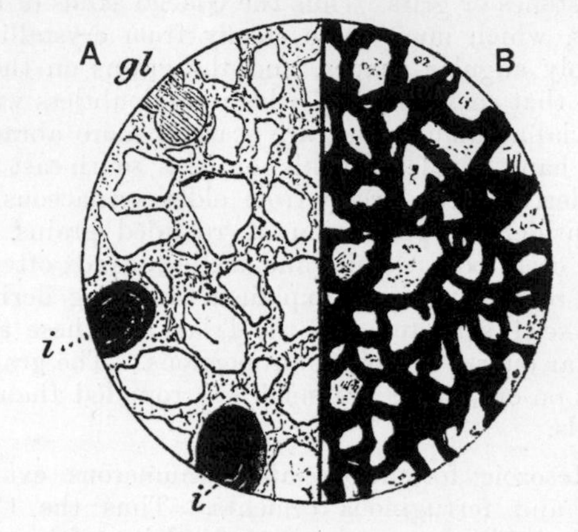

Fig. 50.

A. Calcareous grit in Lower Greensand, Roslyn Hill, Ely; ×20: subangular quartz, with a few glauconite casts of foraminifera (*gl*), and derived oolitic grains of dark brown iron-ore (*i*), cemented by a matrix of granular calcite [1799]. *B*. Ferruginous grit, Kellaways Rock, South Cave, Yorkshire; ×20: shewing angular quartz-grains in a cement mainly of iron-oxide [1797].

The Upper Palæozoic grits and sandstones of this country often have a cement largely ferruginous or consisting of iron-oxide and quartz. In the Devonian of South Devon are fine-grained sandstones which, with predominant quartz, have little flakes of mica, some felspar, and small granules of tourmaline, indicating the source of the material: the interstitial matter is for the most part ferruginous. Much of the Old Red Sandstone shews the investing pellicle of ferric oxide around each grain.

This latter feature and numerous other points of interest may be studied in many parts of the New Red Sandstones. In particular, quartz-grains with a secondary outgrowth having

[1] Hume, *Q. J. G. S.* (1894) l, 679 (Bargate).

crystal-faces are common at various horizons of the Keuper and Bunter[1] of Shropshire, Cheshire, *etc.*, and are also exceptionally well exhibited in some coarse-grained beds of the Penrith Sandstone (Penrith Beacon, Westmorland), (fig. 48). In some cases a pellicle of iron-oxide coats the new crystal-growth, and must then be long posterior to the date of the strata. Red sandstones are often of quite yielding consistency, even when the interstices are occupied by quartz. This is because of the coating of iron-oxide intervening between the interstitial quartz and the original grain. By treatment with acid, the irregularly shaped patches of interstitial quartz were isolated by Mr Phillips from the 'millet-seed' sandstones of the Trias. In these beds the perfectly rounded form of the original grains is attributed to their having been true desert-sands.

The Lower Keuper sandstones at certain localities near Nottingham and elsewhere are cemented by barytes[2]. In different occurrences this exceptional mineral occurs in granular form, in irregular plates enclosing the sand-grains, in good crystals, or in radiate growths (Stapleford and Bramcote Hills and Hemlock Stone).

Many Carboniferous grits have sharply angular grains, and were probably derived directly from crystalline rocks. The coarse-grained Millstone Grit of south Yorkshire[3] has highly irregular quartz-grains poor in fluid-cavities. There is not much fresh felspar, but argillaceous matter between the quartz-grains seems to represent it. The hard 'ganister' has angular quartz-grains which fit so closely together as to obscure the small amount of siliceous cement, and the same is true of the grits of the Bristol coal-field. In some beds in the Coal-Measures numerous flakes of muscovite lying parallel to the lamination impart a fissile character to the rocks (Bradford flags, *etc.*). The spaces between the grains are often obscured by kaolin. Kaolin and relics of reddish orthoclase, with a

[1] For descriptions of Triassic sandstones from the Vale of Clwyd, Cheshire, and Lancashire, see Morton, *Geology of Liverpool* (2nd ed. 1891) 129–132; M. Reade, *Pr. Liv. G. S.* (1892) vi, 374–386.

[2] Clowes, *Rep. Brit. Ass.* for 1885, 1038; 1889, 594; 1893, 732; Watts, *ibid.* 1894, 665, 666.

[3] Sorby, *Proc. Yorks. Geol. Polyt. Soc.* (1859) iii, 669–675.

little mica and sometimes tourmaline are found in the Millstone Grit of south-west Lancashire[1], which consists mainly of angular quartz-grains of very variable size (·2 to ·005 inch) with crystalline outgrowths not very common. In the Cefn-y-Fedw Sandstone of Denbighshire and Flintshire[2] the grains are angular to rounded, and more often have secondary outgrowths with crystal-faces.

The Lower Palæozoic and older arenaceous rocks are as a rule thoroughly compacted, the cement being for the most part siliceous. Mr Phillips found the quartz-cement of various Cambrian and Silurian grits (Barmouth, Harlech, Aberystwith, Denbighshire) permeated by a moss-like growth of a green chloritic mineral. Both coarse and fine-textured rocks are included. The quartz-grains are angular or partly rounded, and frequently contain needles of rutile and tourmaline: fluid-pores are present in some, absent in others. Some of the grits have plenty of felspars, while pyrites, garnet, and micas are occasionally noted. Specimens of the grits of Skiddaw and of the Isle of Man (Santon) shew fragments of slate and lava among the partly rolled quartz and turbid felspars. The Ingleton rock in Yorkshire is a grit containing volcanic material as well as grains and pebbles of quartz, felspars, and various lavas[3]. Volcanic grits of finer texture occur in the upper part of the Ordovician near Shap Wells, Westmorland, and these contain also calcareous matter.

The older sandstones of the Bangor and Caernarvon district and of parts of Anglesey are rather coarse-grained, consisting of well-rounded to subangular quartz with plenty of felspar. The latter mineral is often decomposed and its clayey decomposition-products wedged in between the quartz-grains, obscuring the siliceous cement (fig. 51). Some of the rocks, however, have comparatively fresh felspar: a Silurian grit at Drys-lwyn-isaf, south of Parys Mountain, consists almost wholly of grains of oligoclase closely packed together. The pre-Cambrian Torridon Sandstone is an example of a coarse sandstone rich in felspar. Besides rolled quartz-grains, often

[1] Morton, *Proc. Liverp. G. S.* (1887) v, 280–283.
[2] *Ibid.* 271–279.
[3] Tate, *Rep. Brit. Assoc.* for 1890, 800.

composite, it has others of microcline and fragments of quartzite and pegmatite.

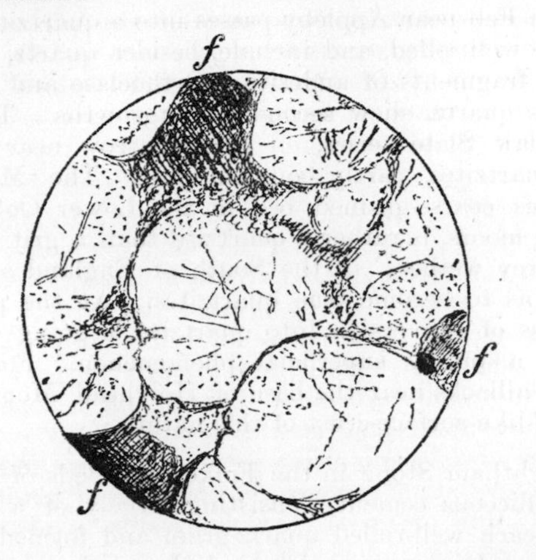

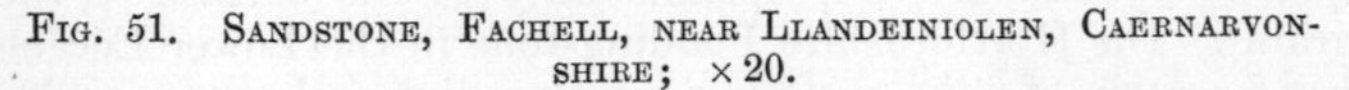
FIG. 51. SANDSTONE, FACHELL, NEAR LLANDEINIOLEN, CAERNARVONSHIRE; ×20.

Besides the well-rolled quartz-grains with many rows of fluid-pores, felspar is represented (*f*). This is largely decomposed, the resulting clayey material being squeezed between the quartz-grains [282].

The best examples of *quartzites* in England are those of Hartshill in Warwickshire and the Lickey Hills in Worcestershire, probably of pre-Cambrian age, and the Stiperstones in Shropshire (Ordovician). All these consist essentially of rolled quartz-grains, usually about ·02 to ·03 inch in diameter, with only very subordinate felspar, united by a clear quartz-cement which is of the nature of a crystalline outgrowth from the grains (fig. 49). A series of quartzites forms the lower part of the Cambrian in the Assynt district, Sutherland. Some beds contain pebbles, and are indeed cemented conglomerates. The uppermost bed ('Top Grit') shews large well-rolled quartz-grains, about ·05 inch in diameter, with smaller subangular grains between them. The remaining space occupied by the siliceous cement is obscured by opaque dust (fig. 47).

Quartzites have been formed in various districts at higher geological horizons. Thus the basal Carboniferous sandstone on Roman Fell near Appleby passes into a quartzite. All the grains are well-rolled, and include, besides quartz, a little felspar, and fragments of andesite. Orthoclase and plagioclase, as well as quartz, shew secondary outgrowths. The grits of the Skiddaw Slate series, on Latterbarrow near Egremont, become quartzites, partly conglomeratic. The 'Moor Grit,' a conspicuous coarse-grained bed in the Lower Oolites of the Yorkshire moors, is rather a quartzite than a grit. The well-known 'grey wethers' in the South of England are often so cemented as to become true quartzites. At the present day the process of conversion into quartzite is going on in some places as a purely superficial phenomenon. For instance, numerous hillocks near the base of Holyhead Mountain shew an enamel-like surface-crust of this nature.

The Ightham Stone in the Folkestone Beds of Kent has a peculiar siliceous cement, consisting largely of a fringe surrounding each well-rolled quartz-grain and formed of minute quartz crystals grown perpendicularly to the surface of the grain[1]. The Hertfordshire Puddingstone in the Lower Eocene has a matrix of flint enclosing angular flint and quartz-grains and large rolled pebbles of flint.

[1] Bonney, *G. M.* 1888, 299.

CHAPTER XVII.

ARGILLACEOUS ROCKS.

THE name *clay* is used for argillaceous deposits which still retain enough moisture to be plastic. By the loss of most of their uncombined water and by other more important changes these pass into mudstones, shales, and slates. Of these terms, *mudstone* is correctly used when the rock has no marked fissile character, *shale* when it splits along the original laminæ of deposition, and *slate* when the original lamination has been superseded as a direction of weak cohesion by a new structure (slaty cleavage, Fr. schistosité, Ger. Transversalschieferung). The Continental geologists do not, as a rule, observe this distinction, but include shales and slates under the same name (Fr. schiste, Ger. Schiefer, Norw. skiffer).

Among slates it is usual to distinguish *clay-slates* (Thonschiefer, lerskiffer), in which the material was supposed to be largely detrital matter without important new formation of minerals, and *phyllites* (Fr. phyllade), in which the rocks are largely or totally reconstituted in place (aided, at least, by pressure). It is now becoming probable, however, that in clay-slates, and even in clays and shales, there has often been a considerable amount of mineral change in place, so that no very sharp line can be drawn between clay-slates and phyllites. The typical glossy phyllites are essentially mica-schists on a small scale, and may be described as micro-crystalline schists. We shall find it convenient to include them here, although we thereby anticipate their place under the head of dynamic metamorphism.

Constituent minerals. Owing to the extremely small dimensions of the elements, it is usually a matter of great difficulty to identify with certainty all the constituents of clays, shales, or slates. Speaking generally, these constituents include some of derived or detrital origin (allothigenous), which were either primary minerals or decomposition-products in the parent rock-masses, and others of secondary origin, formed in place (authigenous). As regards the latter, doubt may exist in particular cases as to how far the secondary recombinations have been induced by pressure (dynamic metamorphism). In many fine-grained slates no constituents are seen which can be set down with confidence as purely detrital. In all cases very thin sections and high magnifying powers must be used. Some of the denser accessory minerals may be isolated from powder by heavy solutions, or merely by washing[1].

The detrital elements may include granules of *quartz*, and less frequently of *felspars*, and scales of *mica*, with minute crystals of such accessories as *zircon*. The little flakes of biotite shew more or less decomposition: Mr Hutchings finds that they give rise especially to *epidote* in minute superposed tablets of light yellow colour. The iron-oxides separate out as *limonite*. *Carbonates* may occur in varying proportion. Many argillaceous rocks contain a considerable quantity of *carbonaceous matter*, finely granular and for the most part opaque: such rocks may be bleached by incineration on platinum foil. The *pyrites* which occurs in many slates, sometimes in relatively large crystals, is of secondary origin, and is perhaps due to the reduction of iron-compounds in the presence of organic matter. The *glauconite* of some argillaceous deposits has also been formed in place.

The ordinary fine-grained argillaceous rocks consist in considerable part of an exceedingly fine-textured base or paste, very difficult to resolve, in which any truly detrital elements or their evident alteration-products are imbedded. The nature of this paste has not yet been made out in any

[1] *Cf.* Teall, *M. M.* (1887) vii, 201–204. For a method of studying fine incoherent sediments, see Hutchings on Sediments dredged from the English Lakes, *G. M.*, 1894, 300–303.

large number of cases. It was formerly regarded as consisting essentially of hydrated silicate of alumina (kaolin, *etc.*). Careful studies of various clays, shales, and slates lead, however, to the conclusion that the material is to a great extent finely divided white *mica*, and it is now generally admitted to be of secondary origin. With this is much indeterminable, finely granular matter, which may be conjectured to represent the finest powder of quartz, felspar, *etc.*, and perhaps *kaolin* or other products. A highly characteristic feature of the paste is the presence of an enormous number of minute needles of *rutile* ('clay-slate-needles'). On account of their very small breadth and very high refractive index, the needles often appear as opaque lines, but the larger ones may be transparent. The rutile is generally regarded as of secondary origin, being produced in place in association with the mica, *etc.*, the titanic acid being furnished by derived biotite. Since the changes which gave rise to these secondary products have operated in clays as well as in slates, they cannot be held to imply any advanced dynamic metamorphism, but they may still have been favoured by pressure.

Many slates seem to shew by their chemical composition the presence of secondary free silica (in addition to any evident detrital quartz which they may contain). This is sometimes seen as a *quartz-cement*, tending to form little veins and patches; in other cases *opal* has been supposed to occur, and indeed amorphous silica may be dissolved out by caustic potash.

In some cases, and more especially in the Glacial tills, we must suppose that a large part of even the most impalpable material is of detrital origin. Thus in the tills of the Boston basin, Massachusetts, Crosby[1] finds that about four-fifths of the finest grade of material is not what is commonly understood by clay, but is what he terms 'rock-flour,' *i.e.* the most minute particles of pulverised quartz and other rock-forming minerals.

Structure. Argillaceous rocks in general have a parallel arrangement of their constituent elements, which is usually

[1] *Proc. Bost. Nat. Hist. Soc.* (1890) xxv, 115–172.

sufficiently marked to impart a fissile character to the mass. Slices parallel and perpendicular to the direction of fissile structure should be compared. In shales a large proportion of the minute constituent elements lie with their flat faces or long axes parallel to the layers of deposition. In true slates, *i.e.* rocks with a superinduced cleavage-structure, they have taken up a new direction along planes (cleavage-planes) perpendicular to the maximum compression by which the rock has been affected.

The effect of this compression, accompanied by a certain partially compensating expansion along the cleavage-planes, is well seen in the deformation of concretionary spots of colour,

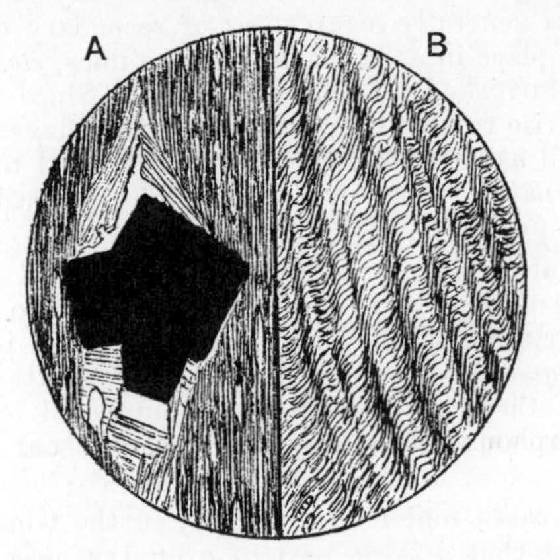

Fig. 52.

A. Slate with crystal of pyrites, Penrhyn, near Bangor; ×5. The crystal is surrounded by an 'eye' of chlorite and quartz, as described. The mass of the slate contains little light spots, which have been deformed into an elliptic shape [501]. *B.* False cleavage in Skiddaw Slate, Brownber, near Appleby; ×20. The system of minute parallel folds causes a direction of weakness equivalent to cleavage [913].

etc. A spherical spot becomes distorted into an ellipsoid. A hard unyielding body, such as a crystal of pyrites or magnetite imbedded in the rock, gives rise to curious phenomena. The matrix flows past the crystal, leaving a roughly eye-shaped space[1]. Such crystals have in many cases been originally coated with an envelope of chlorite, which adheres to the matrix and is torn away from the crystal. The intervening space is subsequently filled by infiltration with crystalline quartz (fig. 52, *A*).

Various structures, of frequent though local occurrence in fine-grained beds, may be styled false and incipient cleavages[2]. They consist sometimes in a parallel system of microscopic faults, sometimes in a regular system of minute folds. These often give a tendency to the rock to split along definite planes, *viz.* the fault-surfaces or the limbs of the folds (fig. 52, *B*). Dr Sorby[3] has shewn that such structures may be a step towards a true slaty cleavage. They may also, however, occur as later structures crossing a true cleavage (*e.g.* in various Ardennais slates and phyllites), and they are common in some fine-textured mica-schists. They are often interesting as reproducing on a minute scale the characteristic structures of mountain-ranges, such as the gradual passage of an over-fold into an overthrust fault, the relation of faults to anticlines, *etc.* A frequent result of shearing movement in finely laminated rocks is the formation of minute oblique folds inclined at about 45° to the lamination: these are pushed over until at about 30° they pass into little faults, and the faults may be further pushed over until they are lost in a general parallel-structure.

Illustrative examples. Before describing some of the commoner types of argillaceous rocks, we may mention one of which very little is known among consolidated strata. It is represented among deposits now forming by the *abyssal red clay* which covers large areas of the ocean-floor below a depth of 2200 fathoms. This deep-sea clay is derived mainly from the destruction of volcanic products by the chemical action of sea-water. Minute fragments of volcanic rocks and minerals are mixed with decomposition-products and with a few siliceous organisms (radiolarians, *etc.*). The brownish red colour is due to disseminated limonite. Minute crystals of the lime-zeolite

[1] *G. M.*, 1889, 396, 397.
[2] *Rep. Brit. Ass.* for 1885, 836–841.
[3] *Q. J. G. S.* (1880) xxxvi, *Proc.* 72, 73.

phillipsite or christianite are common[1], and manganese-nodules of various sizes occur. There may be a few corroded tests of foraminifera. A rock comparable with deep-sea clay was noted by Dr Guppy[2], among the Recent deposits of the Solomon Islands in the Pacific. Some grey mudstones and fine clays from Barbados[3] also resemble the 'red clay.'

These deep-sea argillaceous deposits have characters which distinguish them from those derived from the waste of land-areas. The particles are of excessive minuteness and markedly angular in shape[4]. The minerals recognizable are those most common as constituents of volcanic rocks, such as felspar and augite, rarely quartz, while such minerals as zircon, tourmaline, *etc.*, are absent. Usually a very large proportion of the material consists of angular chips of volcanic glass and elongated fragments derived from the breaking up of pumice with capillary pores.

As another somewhat peculiar type of clay may be mentioned the *china-clay* of Cornwall, which seems to consist essentially of the mineral kaolin[5]. This, in its most recognizable form[6], builds minute colourless scales, sometimes with hexagonal outline, and of such refractive index and birefringence as closely to resemble mica. It appears, however, from Mr Collins' account[7] that these distinct flakes do not form any large part of the finely divided material in the typical occurrences in Cornwall. Besides quartz, mica, and other impurities, tourmaline is found in some rocks composed largely of kaolin, and its production was perhaps connected with the process of 'kaolinization' of felspathic rocks[8]. In addition to

[1] Murray and Renard, Challenger Report (1891), *Deep-Sea Deposits*, Pl. XXII.

[2] *The Solomon Islands, their Geology, etc.* (1887), 81, 82.

[3] Miss Raisin, *Q. J. G. S.* (1892) xlviii, 180–182; see also Gregory, *ibid.*, 539 (Trinidad).

[4] Murray and Renard, *l. c.*, Pl. XXVI, XXVII, figs. 1–4; contrast with fig. 5.

[5] Some writers apply the name kaolin to the clay itself, and use 'kaolinite' for the mineral.

[6] See Dick, *M. M.* (1888) viii, 15–27, Pl. III.

[7] *M. M.* (1887) vii, 205–214.

[8] Butler, *M. M.* (1887) vii, 79, 80; *etc.*

the proper china-clays, formed more or less *in sitû*, there are derived clays of similar composition, such as those of Bovey Tracey.

In contrast with the preceding types of clays, in which the substance is derived almost wholly from the chemical degradation of preëxisting rocks, may be mentioned glacial *till*, the material of true ground-moraines, in which, as already noticed, a very large proportion of the mass is directly due to the mechanical grinding down of solid rocks. The petrology of the finer glacial accumulations (apart from their contained boulders) is, however, an almost untouched subject.

We pass on to the consideration of clays and slates of more ordinary constitution, selecting only a few examples which may be regarded as typical.

A minute study of typical argillaceous rocks has been made by Mr Hutchings[1] in the case of the *fire-clays* of the Newcastle Coal-measures. The rocks are laminated, and include coarser and finer beds. The material of true detrital origin is most abundant in the coarser beds. It seems to be derived from the destruction of granite, and consists of granules of quartz averaging ·002 to ·003 inch in diameter, granules of felspar, biotite flakes from ·01 inch downward, with the epidotic alteration, less abundant muscovite, and accessory zircon, *etc.* Besides these there is a paste, in which minute scales of secondary mica and needles of rutile are the recognizable elements.

The Culm-measure *shales* of Bude in Cornwall[2] are derived from the waste of granite (in part with tourmaline) and crystalline schists. They appear to have undergone more change *in sitû* than the preceding.

The Cambrian *roofing-slates* of North Wales represent a more advanced stage of secondary change, both structural and mineralogical. They possess a strong cleavage-structure, passing indifferently through the layers of original deposition, and the more altered of them have the glossy aspect of fine-textured phyllites, in which little trace of any clastic structure survives.

[1] *G. M.* 1890, 264–273.

[2] McMahon, *G. M.*, 1890, 108–113; Hutchings, *ibid.* 188.

Detrital granules of quartz and felspar may be seen, but biotite is wanting, though little patches of epidote perhaps represent it. "The base and main constituent of all these slates is a fine-grained mica, mostly lying flat in the plane of cleavage of the rock," and rutile-needles are usually abundant. The red and purple slates contain numerous scales of red micaceous hæmatite, probably representing the limonite of less altered deposits.

The Devonian slates of Cornwall (Tintagel, *etc.*) are described by Mr Hutchings[1] as having suffered more alteration (ascribed to dynamic metamorphism) than the Welsh rocks. They have no clastic quartz, felspar, or biotite, and indeed some very small zircons seem to be the only derived constituents left unaltered. The main mass of the rock is of fine sericitic mica, the majority of the minute flakes being parallel to the cleavage of the rock. Minute needles of rutile are very abundant. Another very common mineral is micaceous ilmenite in flakes about ·002 inch in diameter. This is either opaque or transparent, with a deep brown colour, and sometimes encloses characteristic skeletons of rutile (sagenite). Other constituents of some of these slates are secondary quartz, calcite, chlorite, ottrelite, garnet, *etc.*

The Cambrian *phyllites* of the Ardenne have been carefully examined by Renard[2], who finds that the rocks have been completely reconstituted in place. The chief mineral is usually a colourless sericitic mica, its flakes having a general parallelism with the cleavage or schistosity of the rock. This and quartz usually constitute the principal part of the bulk, and a green chlorite is also abundant. Needles of rutile and often of tourmaline lie in general parallel to the cleavage. The violet phyllites have micaceous hæmatite ('oligiste'); in others micaceous ilmenite occurs, with interpositions of sagenite. Other minerals found in particular rocks are magnetite and pyrites, a manganese-garnet (spessartine) in minute crystals, ottrelite, zircon, carbonaceous matter, *etc.* The magnetite in the 'phyllade aimantifère' was formed before the cleavage of the rock, and is surrounded by the curious eyes of chlorite and

[1] *G. M.* 1889, 214–220; *ibid.* 1890, 317–320.
[2] *G. M.* 1883, 322–324 (*Abstr.*).

quartz already referred to. The ottrelite was formed subsequently to the cleavage of the rocks which contain it, and its flakes do not lie parallel to the cleavage-planes.

Of ordinary slaty cleavage good illustrations are afforded by the Cambrian and Ordovician in North Wales, the Devonian in Cornwall, and some other British Palæozoic rocks. Some of these (Llanberis Slates) exhibit the deformation of originally spherical spots. Various kinds of 'eyes' about enclosed pyrites crystals may be seen at Penrhyn (fig. 52, *A*), Snowdon, Blaenau Ffestiniog, Whitesand Bay, *etc.* Special structures of the nature of *false cleavage* may be examined at particular places in the Snowdon district (Drws-y-coed Pass), in the Skiddaw Slates of the Eden valley (Brownber, near Appleby[1], fig. 52, *B*), and of Snaefell in the Isle of Man, in the debatable rocks of the Start in South Devon[2], and in the remarkable 'gnarled' beds of Amlwch in Anglesey and of Aberdaron, *etc.*, in the west of Caernarvonshire. These last shew very beautifully all the characteristic structures of 'mountain-building' on a small scale, and such rocks afford from this point of view an interesting study. Prof. Heim, in a figure[3] illustrating the passage of an overfold into an overfault by the obliteration of the 'middle limb,' gives for the scale '$\frac{200}{1}$ to $\frac{1}{10000}$ of natural size.' Perhaps the best district for studying the various forms of false cleavage is the Isle of Man, where the Skiddaw Slates exhibit a great variety of interesting structures.

[1] *Q. J. G. S.* (1891) xlvii, 513, 514.
[2] *G. M.*, 1892, 347, 348, Pl. VIII, fig. 1.
[3] *Mechanismus der Gebirgsbildung* (1883), Pl. xv, fig. 14.

CHAPTER XVIII.

CALCAREOUS ROCKS.

THE different kinds of limestones (Fr. calcaire, Ger. Kalkstein), consisting of carbonate of lime with various impurities or foreign material, are almost all in great measure of organic origin. The hard parts of calcareous organisms are composed of calcite or aragonite, or both, with a small quantity of phosphate, *etc.* It will be seen that aragonite is always the unstable form of carbonate of lime, and tends to be converted into the stable form, calcite.

The impure calcareous rocks may include a considerable amount of non-calcareous material; either sand-grains (calcareous grit) or finer detritus (argillaceous limestone, marl) or volcanic *dêbris* (calcareous tuff).

With the limestones must be classed those rocks in which dolomite takes the place of calcite. These are called dolomite-rocks or dolomites, the name dolomitic limestone or magnesian limestone being more correctly applied to rocks in which both minerals are well represented. Many dolomitic rocks can be proved to have originated from ordinary limestones, the magnesia which replaced part of the lime having been derived from some external source. In the view very widely held this process of replacement is considered to be the most important factor in the origin of magnesian limestones and dolomite-rocks in general; but it is by no means certain that the magnesia is always thus introduced from without. Most calcareous organisms contain a small amount of carbonate of magnesia (less than 1 per cent.), and in some rock-forming calcareous algæ (*Lithothamnion*) the proportion is as much as 10 per cent. It is probable that this is secreted in the form of dolomite. There is evidence that, in consequence of calcite being much more readily soluble than dolomite, the proportion of magnesia to lime may be considerably raised[1], especially in detrital calcareous mud and in finely divided sediment which has remained suspended in water for some time.

We shall also briefly notice certain other rocks, such as some bedded iron-stones, which are genetically connected with the limestones, and some siliceous deposits.

Much valuable information concerning limestones is contained in Dr Sorby's Presidential Address to the Geological Society[2], while British limestones from various horizons have been studied by several other observers[3].

Organic fragments. Most of the fragments of calcareous organisms that form part of rocks have something in their mineral nature, their structure, or their mode of preservation, that enables us to refer them to their proper order or class, or at least sub-kingdom.

Among vegetable organisms, the *calcareous algæ* figure largely in the deposits now forming around coral-islands[4] and to a less extent in some deep-sea deposits, while the equivalents of these rocks are recognized among the Tertiary and Recent strata in various parts of the world[5]; *e.g.* the Lithothamnion

[1] See *e.g.* Hardman on Irish dolomites, *Proc. Roy. Irish Acad.* (1877), ser. 2, ii, 705–730, for a discussion of the relative solubility of the carbonates.

[2] *Q. J. G. S.* (1879) xxxv, *Proc.* 56–95. On calcite and aragonite organisms, see also Cornish and Kendall, *G. M.*, 1888, 66–73. Also Nicholson and Lydekker, *Manual of Palæontology*, 3rd ed. pp. 17–31, *etc.*

[3] See especially several papers by Wethered, *Q. J. G. S.* (1888–1893) xliv-xlix, *etc.* Jukes-Browne and Hill on Chalk, *etc.*, *ibid.* (1887–1889) xliii–xlv.

[4] See Murray and Renard, Challenger Report on *Deep-Sea Deposits* (1891), Pl. XIII, XIV.

[5] Nicholson and Lydekker, p. 24, fig. 9 (Leitha); Murray, *Scott. Geog. Mag.* (1890) vi, Pl. I (Malta); Hill, *Q. J. G. S.* (1891) xlvii, 243–248, Pl. XII (Barbados); Lister (and Murray), *ibid.* 602, 603 (Tonga Is.); Gregory, *Q. J. G. S.* (1892) xlviii, 538–540 (Trinidad); Hinde, *Q. J. G. S.* (1893) xlix, 230, 231 (New Hebrides).

Limestone and Leitha Limestone of the Vienna basin (compare fig. 53). Calcareous algæ are concerned in the formation of some modern oolitic accumulations, and Girvanella, which figures largely in association with oolitic structure in rocks of various ages, is perhaps a vegetable organism; while the

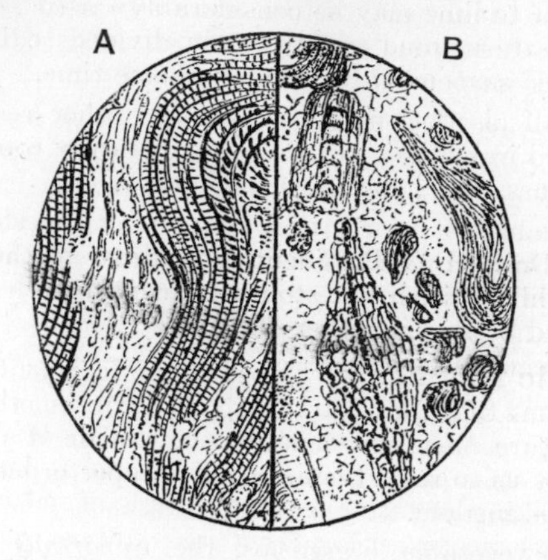

FIG. 53. RECENT ORGANIC LIMESTONES, COMPOSED LARGELY OF CALCAREOUS ALGÆ, EUA, TONGA ISLANDS; × 20.

A is a characteristic section of *Lithothamnion* [1271]. *B* shews foraminifera and fragments of algæ in a recrystallized calcareous matrix [1269].

peculiar algous flora of hot springs is instrumental at the present day in producing certain deposits of travertine (Mammoth Hot Springs[1]). The part played by algæ in the formation of some of the older limestones, such as the Alpine Trias, seems to be of considerable importance[2].

The tests of calcareous *foraminifera* commonly occur entire, and are readily recognized, though in some cases the chambers become detached (Globigerina). The material is calcite or aragonite in different forms (answering to the division into Vitrea and Porcellanea of some authors), and probably the

[1] Weed in 9th *Ann. Rep. U. S. Geol. Surv.* (1890), 642–645, *etc.*
[2] Seward, *Science Progress* (1894) ii, 10–26.

latter have been largely destroyed in some older limestones. Foraminifera occur in many shallow-water limestones[1], and make up a large part of the so-called coral-limestones[2], besides forming the bulk of extensive deep-sea deposits. The Nummulitic Limestone is a well-known instance of a rock composed largely of foraminifera. Other examples are the Alveolina or Miliolite Limestone of Mixen Rocks near Selsea and the Saccamina Limestone of Northumberland[3].

The interior of a foraminiferal test may be filled in by crystalline calcite, often with such a radial arrangement of fibres as to give a very perfect black cross in each chamber when examined between crossed nicols. In many modern sediments[4] formed near a continental shore-line the chambers are occupied by a deposit of green *glauconite*, which, by the removal of the calcareous test, may be left in the form of casts, and this seems to be the usual mode of origin of glauconite-sands, such as are found at various geological horizons[5].

The true *corals* consist, according to Dr Sorby, of little fibres, or in some cases granules, of aragonite, but it appears that calcite enters into the composition of some forms. Of the Rugosa some consist largely of calcite fibres roughly parallel to the outlines of the several parts of the skeleton, while the mode of preservation of others seems to indicate that they were composed largely of aragonite. The so-called coral-rock, coral-sand, and coral-mud of Recent strata and of deposits now forming often consist largely of calcareous algæ or foraminiferal tests, but some are of almost pure corals and coral fragments. Among older rocks having this constitution may

[1] *E.g.* Guppy, *Tr. Roy. Soc. Edin.* (1885) xxxii, Pl. CXLV, figs. 1, 4 (Solomon Is.); Jennings, *G. M.*, 1888, Pl. XIV (Orbitoidal Limestone of Borneo).

[2] See Guppy, *The Solomon Islands, Geology, etc.* (1887), 73–76; and *Tr. Roy. Soc. Edin.* (1885) xxxii, 545–581; Lister (and Murray), *Q. J. G. S.* (1891), xlvii, 602–604.

[3] Nicholson and Lydekker, p. 126, fig. 30. On the inorganic nature of the so-called Eozoon see *ibid.* pp. 137–143.

[4] Murray and Renard, Challenger Report, *Deep-Sea Deposits* (1891), Pl. XXIV, XXV.

[5] See, *e.g.*, Murray, *Scott. Geog. Mag.* (1890) vi, 464, 465, Pl. II, fig. 2 (Malta); Gregory, *Q. J. G. S.* (1892) xlviii, 540 (Trinidad).

be mentioned parts of the Mountain Limestone and the Coral Oolite and certain Devonian limestones of South Devon.

The hard parts of *echinoderms* have an unmistakable appearance. Each element (plate or joint) behaves optically as a single crystal of calcite, the larger ones shewing the characteristic cleavage[1]. The organic nature is indicated only by the external form, internal canals, *etc.* Spines of echinoids, joints of the stems of crinoids, *etc.*, may be distinguished by their size and outline (fig. 54).

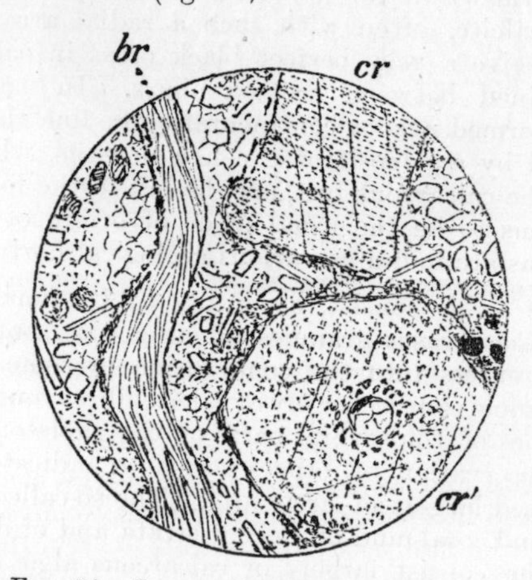

FIG. 54. LIASSIC LIMESTONE, SKYE; ×15: shewing joints of crinoid stems (*Pentacrinus*) cut longitudinally (*cr*), and transversely (*cr′*), each consisting of a single crystal of calcite; also part of a brachiopod shell (*Rhynchonella*, *br*), with its characteristic fibrous structure. The matrix is a recrystallized calcite mosaic enclosing numerous detrital grains of quartz and flakes of muscovite [1791].

The structure of the hard parts of *crustacea* is also fairly constant and quite different from the preceding. The shell is built of fibres of calcite set everywhere perpendicular to the surface, the optic axis of each fibre coinciding with its length. The general outline suffices to distinguish, *e.g.*, between ento-

[1] Nicholson and Lydekker, p. 23, fig. 7. For minute structure of plates, see *ibid.* p. 364, fig. 236.

mostracan tests (abundant in many limestones) and fragments of trilobites (fig. 55, *A*).

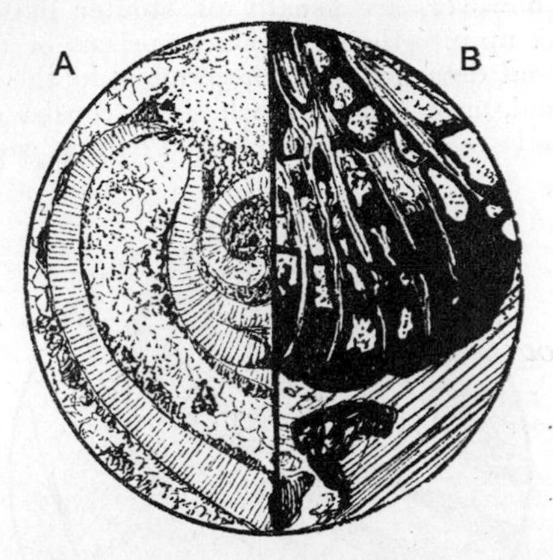

FIG. 55. CARBONIFEROUS LIMESTONE, CLIFTON, BRISTOL; ×20.
A shews a portion of a trilobite with the characteristic structure of the crustacea [981]. *B* polyzoa replaced by opaque limonite, mixed with silica, in a matrix of conversely crystalline calcite [972].

Both calcite and aragonite enter into the composition of the *polyzoa*, and in some genera, according to Messrs Cornish and Kendall, the two occur in separate layers, the aragonite layer being in this case the outer one.

The shells of *brachiopods* are wholly of calcite, with a characteristic structure. "They are made up of laminæ, consisting of flattened fibres or prisms, often passing along more or less parallel to one another over a considerable area, but mixed up with other systems which cross them at various angles." These laminæ lie oblique to the surface of the shell, and the individual fibres do not give strictly straight extinction (fig. 54). The 'perforations' of some brachiopod shells can be seen, but from Dr Carpenter's investigations it seems doubtful whether this character is of more than specific value.

The shells of *lamellibranchs* have more than one type of structure. In some ostreid genera (Ostrea, Pecten, Gryphæa,

Inoceramus) the whole is of calcite in irregular flattened fibrous plates, producing a structure not unlike that of brachiopods. The shells, however, are usually of stouter build, and they tend to break up into their component prisms or fibres, which are often found detached, *e.g.* Inoceramus in the Chalk. On the other hand, most lamellibranch shells consist originally of aragonite, and are commonly preserved only as casts in calcite

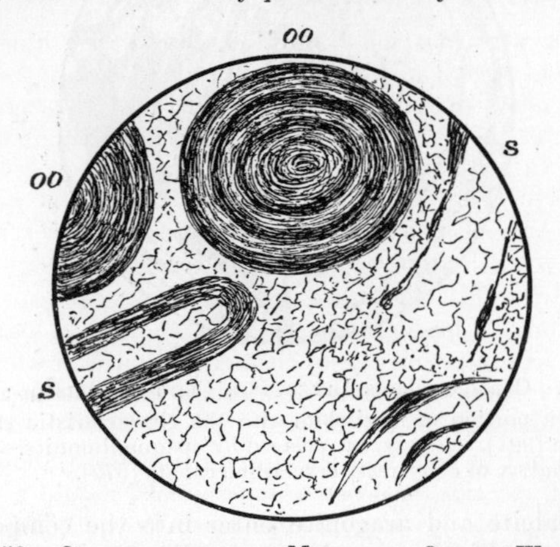

FIG. 56. OOLITIC LIMESTONE, MILLEPORE OOLITE, WHARRAM, YORKSHIRE; ×20: shewing oolitic grains (*oo*) and chips of lamellibranch shells (*s*) in a matrix which has recrystallized as a mosaic of clear calcite [1794].

mosaic (fig. 56). In some genera (Pinna, Mytilus, Spondylus) there is, according to Dr Sorby, an inner layer of aragonite protected by an outer layer of calcite.

Most *gasteropods* have shells wholly of aragonite, which is readily replaced by a mosaic of crystalline calcite. In some cases, however, *e.g.* Scalaria, the whole is of calcite (Cornish and Kendall). Others have a layer of aragonite covered by a layer of calcite: either the former (Murex) or the latter (Purpura) may form the bulk of the shell.

Of the *cephalopoda*, the shells of Nautilus and ammonites were originally of aragonite, but the aptychi of the ammonites were calcite. The belemnites had the guard of calcite, with a characteristic radial arrangement of fibres about an axis, but the phragmocone was aragonite.

The tests of *pteropoda* may sometimes be recognized by their form in sections. Exceptionally they form the main constituent of a limestone[1], and 'pteropod ooze' is one of the deep-sea deposits now forming in some parts of the ocean.

Oolitic structure[2]. Many shallow-water limestones, of all geological ages, contain little spheroidal grains built up of successive coats of calcareous material, and these may be so numerous as to make up the chief bulk of the rock. Such rocks are called oolitic limestones, oolites, or roestone (Ger. Rogenstein). For the coarser types, in which the grains may reach the size of peas, and are often of rather irregular or flattened form, the name pisolite (Ger. Erbsenstein) is used.

In addition to the concentric-shell arrangement, there is often a more or less evident radial structure in each grain[3], and closer examination shews that the minute elements which build up the successive layers are set in some cases radially, in other cases parallel to the layers.

As a result of either of these arrangements an oolitic grain, examined in section between crossed nicols, should give a black cross comparable with that observed in the spherulites of igneous rocks[4]. Owing to the departure from true sphericity, the admixture of granular material not sharing the definite orientation described, and the effect of iron-staining and other secondary changes, an accurate black cross is not seen in every case.

The concentric layers have been formed upon a nucleus, which may be a chip of shell or other organic body, a quartz-granule, or merely a pellet of fine calcareous mud. Similar coatings are often to be seen upon fragments of shell, *etc.*, too

[1] Nicholson and Lydekker, p. 24, fig. 8.

[2] On the oolitic structure and its significance see Sorby's *Presid. Address, l. c.*; also Teall in *Mem. Geol. Surv., Jurassic Rocks of Britain*, vol. iv, pp. 8–12, Pl. I, II, 1894; Wethered (papers cited).

[3] See, *e.g.*, Nicholson and Lydekker, p. 28, fig. 11.

[4] Rosenbusch-Iddings, Pl. IX, fig. 2.

large to be built up into round grains. Sometimes an oolitic grain has been broken and the separated fragments subsequently coated with fresh layers of calcareous deposit; or again two or three contiguous grains may be enveloped in one mantle and become a compound grain.

Oolitic grains differ as regards their material (calcite or aragonite), the orientation of their minute elements (radial or tangential), the presence or absence of finely granular calcareous matter without special orientation, or of impurities, and in other respects. One common type, exemplified in many British limestones, has well-marked concentric shells, each of which consists largely of minute calcite prisms or fibres set radially. There may or may not be an evident radial structure in the grain as seen in a thin slice. The black cross seen in polarized light is often imperfect or vaguely defined.

Another type is illustrated by the so-called Sprudelstein of the Carlsbad hot springs. Here there are well-marked concentric shells but no radial structure. The material is aragonite, and the minute elements are set mainly tangentially to the concentric layers. This gives a well-defined black cross. Dr Sorby found recent oolites from Bahama and Bermuda to have a similar constitution, but with some unoriented granular material, and he observed the same in the Bembridge Limestone of the Isle of Wight.

It is impossible to say with certainty to what extent aragonite oolitic grains have once been represented in our older rocks. In numerous instances the present structure of the grains shews that they have been recrystallized. They often consist of crystalline calcite, either in a mosaic or in wedges with a rough radial arrangement. In some cases there is an eccentric radial structure, as if the recrystallization had started at one or more points on the circumference of the grains.

It has been a somewhat difficult question how far the original structure of the different types of oolitic grains is due on the one hand to mechanical aggregation or on the other to crystallization, and it is now being suggested that organic agency may often have played an important part. The Carlsbad Sprudelstein, the modern calcareous sand of Salt Lake, and other oolites seem to be connected with lime-secreting algæ, while Mr Wethered[1] finds the problematical organism

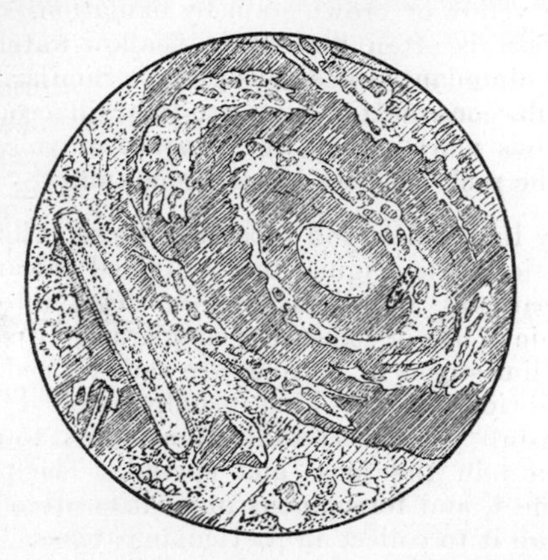

FIG. 57. PISOLITIC GRAIN FROM THE PEA GRIT OF THE INFERIOR OOLITE, LECKHAMPTON, NEAR CHELTENHAM; ×10.

The concentric coats are built up largely of the interlacing tubes of *Girvanella*, seen in cross-section in numerous places: the nucleus is a fragment of a crinoid [1587].

Girvanella in many oolitic rocks of various ages. It is well seen encrusting the successive layers of large pisolitic grains in such a rock as the Pea Grit of Cheltenham (fig. 57).

Matrix of limestones. Recognizable fragments of organisms, together with oolitic grains, if present, may make up a variable part or even the chief bulk of a limestone. The remainder, in rocks which have suffered no important secondary changes, consists of a calcareous mud in which the fragments (and oolitic grains) are imbedded. This finely divided material is mostly carbonate of lime, and must be in great measure derived from the attrition and disintegration of calcareous

[1] See papers cited below, but especially *Q. J. G. S.* (1895) li, where the organic theory is extended to oolitic limestones in general.

organisms, though chemical deposition may perhaps play some part, and material may be furnished by the degradation of older limestones. Iron-compounds often occur as an impurity, producing a yellow or brown stain by oxidation. Fine sand of detrital origin is often present in shallow-water limestones, and may be abundant (calcareous grits). Similarly, an admixture of argillaceous matter gives rise to argillaceous limestones and calcareous marls, or by the presence of volcanic detritus and ashes the rock becomes a calcareous tuff.

In many limestones, and especially those belonging to the older formations, the original finely divided calcareous matter has been partially or wholly *recrystallized* into a granular calcite-mosaic of fine or sometimes comparatively coarse texture. Crystalline limestones or marbles are thus formed without any special conditions of the kind usually implied in the term metamorphism. The recrystallization seems to originate at certain points in the mass and spread. The process has a purifying effect, and ferruginous impurities often appear as if pushed before it to collect in particular patches. The recrystallized carbonate of lime is always calcite, aragonite being converted in the process to the stabler form. In such a crystalline matrix casts after aragonite shells may usually be recognized by a rather coarser mosaic and by a thin film of impurities marking the original outline, even when they are not coated in oolitic fashion (fig. 56).

The recrystallized calcite usually forms a more or less finely granular mosaic in the interstices between the organic fragments, oolitic grains, *etc.* In some cases, however, the individual crystal-grains of calcite are of large size, so as to enclose numerous oolitic granules, shell-fragments, *etc.*, thus giving a structure like the ophitic and pœcilitic in some igneous rocks. This has been remarked by Mr Teall in some of the oolitic building-stones of the Lincolnshire Limestone (Barnack, Ketton, Ancaster). An analogous structure has already been noted above in certain calcareous grits with abundant calcite matrix, the Fontainebleau Sandstone affording an extreme example.

In certain coarse-textured marbles the new-formed calcite occurs partly as a crystal outgrowth of fragments of crinoids, *etc.*, comparable with the quartz-cement of many quartzites (Clifton). On the other hand the crystallization of the matrix may extend to the enclosed fragments, so that the cleavage-planes of the calcite pass continuously from one to the other: this, at least, is the interpretation given in the case of the Keisley Limestone in Westmorland[1].

In some oolitic limestones the original matrix has been in great measure removed by solution, leaving vacant spaces between the oolitic grains. This is seen in some of the Ancaster and Ketton building-stones, belonging to the Lincolnshire Limestone[2].

Although a finely divided matrix and any aragonite organisms present are the parts most readily transformed by these secondary actions, the whole mass of the limestone may in some cases lose every trace of original structure, passing into a compact or granular mass. According to Dr Walther, the accumulations built up by calcareous algæ are peculiarly liable to be changed into 'structureless' limestone.

The quartz-sand, *etc.*, occurring as impurities in many limestones can be easily isolated by dissolving the rock in dilute acid, and sometimes present points of interest[3]. Minute perfect crystals of quartz may occur, sometimes evidently formed by secondary outgrowth from detrital quartz-grains (Clifton).

Deep-sea calcareous deposits. Beyond the broad belt of deposits now forming along the continental coast-lines and deriving their material in some degree from the waste of the land and from shallow-water organisms, and apart too from the special accumulations forming round coral- and volcanic islands, extensive calcareous deposits are found covering large areas of the floor of the deep ocean down to about 2800 fathoms. The most widely spread of these deposits is *globigerina-ooze*,

[1] Nicholson and Lydekker, p. 20, fig. 5 *A*.

[2] On this and some other north-country Jurassic limestones, see *Naturalist*, 1890, 300–304.

[3] Wethered, Carboniferous, *Q. J. G. S.* (1888) xliv, 186–198; Inferior Oolite, *ibid.* (1891) xlvii, 559–569.

consisting largely of the tests of Globigerina and other foraminifera[1], together with a smaller proportion of other organisms, such as siliceous radiolaria, and some non-calcareous matter of volcanic origin. Associated with the foraminiferal remains are immense numbers of very minute elliptic disc-shaped bodies, to which Prof. Huxley gave the name *coccoliths*[2]. These calcareous discs have been detached from the surface of certain globular organisms named coccospheres, referred to the algæ. The coccoliths have a diameter of ·0002 to ·0005 inch. Associated with them are often other minute bodies in the form of slender rods with a crutch-like termination (rhabdoliths). Coccoliths and rhabdoliths are very characteristic of the deep-sea calcareous deposits, though not confined to them.

The inorganic residue of these rocks is essentially of volcanic material in a state of extremely fine division, and corresponds with the 'red clay' already noticed (p. 211).

Various foraminiferal and other limestones have been described among Tertiary and Recent strata which approximate, in some cases very closely, to the essential characters of true deep-sea deposits[3].

Metasomatic changes in limestones. In many rocks which may be assumed to have been once ordinary limestones, the carbonate of lime has been partly, or even wholly, replaced by other substances, thus producing a change in the chemical composition of the rock (metasomatism). The most common of such changes is that in which calcite is converted into dolomite by the replacement of half its lime by magnesia (*dolomitization*). It seems to be clearly established that calcite and dolomite are not chemically isomorphous substances, but each has its own definite composition. The molecular ratio CaO : MgO in dolomite is always unity, and a higher ratio in

[1] Murray and Renard, Challenger Reports, *Deep-Sea Deposits* (1891), Pl. XI, figs. 1, 5, 6; XII; XV, fig. 2.

[2] *Ibid.*, Pl. XI, figs. 3, 4. See also Wallich, *Ann. Mag. Nat. Hist.* (1861) 3, viii, 52–56, and on the coccoliths of the Chalk see Sorby, *ibid.*, 193–200.

[3] *E.g.* Hill, *Q. J. G. S.* (1892) xlviii, 179 (Barbados).

the bulk-analysis of a dolomitic rock indicates a mixture of dolomite and calcite.

In the finely granular mosaic which such rocks often present it may be difficult to distinguish the two minerals from one another without chemical tests[1]. One criterion is the much stronger tendency of dolomite to develope crystal

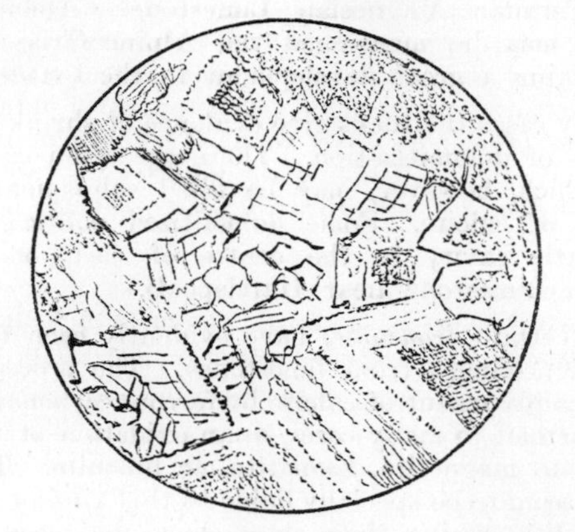

FIG. 58. DOLOMITIZED LIMESTONE IN UPPER CONISTON LIMESTONE, SHAP WELLS, WESTMORLAND; ×20.

The dolomite is here in good rhombohedra with a zonary structure marked by inclusions: some calcite remains as a clear mosaic [1616].

outlines, always those of the primitive rhombohedron (fig. 58). In coarse-grained rocks the more marked cleavage-traces of calcite and the frequency in it of lamellar twinning help to distinguish it from dolomite. Again, calcite is colourless in slices, while dolomite usually (but not always) has a yellow or yellowish brown tint. This coloration is probably due to iron. It may be remarked that another mineral of the same group is sometimes met with, *viz.*—chalybite or siderite, the ferrous carbonate. This often builds little rhombs with curved outlines. It is of a somewhat deeper brown tint than dolomite,

[1] Lemberg has given a microchemical test applicable to rock-slices; *M. M.* (1889) viii, 166 (*Abstr.*).

and in many cases encloses little opaque specks or minute crystals of pyrites.

Good examples of more or less perfectly dolomitized rocks occur in the Durness Limestones of Sutherland, the Bala and Coniston Limestones, the Devonian of Devonshire, the Carboniferous Limestone of many parts of England and Ireland, and the Permian Magnesian Limestone. Among foreign formations may be mentioned the Alpine Trias, dolomitic rocks attaining a great development in the southern Tirol.

In many cases the rocks give evidence of shrinkage during the process of dolomitization. There are often crevices and cavities, which, however, may be filled subsequently by an infiltration of calcite. Some dolomitized oolitic limestones shew a little cavity in the centre of each oolitic grain (Magnesian Limestone near Hartlepool).

Again, certain *ironstones* have evidently been formed[1] by metasomatic changes from limestones. The process consists first in the replacement of calcite by ferrous carbonate (chalybite), and further, in many cases, in an oxidation of the latter, giving rise to magnetite, hæmatite, or limonite. The oolitic limestones seem to be specially liable to this kind of alteration, and the oolitic grains themselves shew the most advanced stage, the outer part of each grain being converted into magnetite or limonite, while the matrix of the rock remains as chalybite or in part calcite. The chalybite matrix is fine-textured, and the mineral often shews imperfect crystal form, each crystal sometimes enclosing a nucleus of decomposing pyrites (fig. 59, *A*). In a more advanced stage of change patches of limonite replace the chalybite of the matrix (fig. 59, *B*), and even calcite shells of Pecten, *etc.*, are converted into hæmatite or limonite (*e.g.* the Dogger of the Peak in Yorkshire). The oxidation does not take place in the more argillaceous ironstones, the iron remaining there in the form of carbonate. Valuable oolitic ironstones are worked in this country. That

[1] See Sorby, *l.c.* pp. 54, 55; Judd, *Geol. of Rutland*, 117–138; Hudleston, *Proc. Geol. Ass.* (1889) xi, 117–127; Cole and Jennings, *Q. J. G. S.* (1889) xlv, 426, 427; Teall in *Mem. Geol. Surv., Jurassic Rocks of Britain*, vol. iv, Pl. II, *etc.*

of Rosedale (Dogger) is magnetite, the Cleveland Main Seam (Middle Lias) shews various stages of transformation and various admixtures of earthy matter, the Jurassic ores of Northampton and Rutland have specially the limonite type

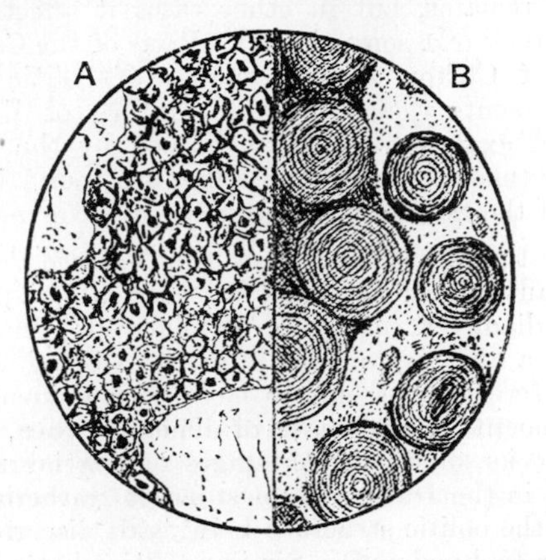

FIG. 59. IRONSTONES.

A. Ironstone-band in Scarborough Limestone, Scarborough; ×100, shewing an aggregate of minute rhombs of chalybite, often enclosing nuclei of pyrites. The clear grains, of which two are shewn, are quartz [946]. *B.* Oolitic ironstone, Claxby, Lincolnshire; ×20. Here the oolitic grains are transformed to limonite: the matrix is mostly of chalybite, but has undergone in patches the further change to limonite [1591].

of alteration, and the Neocomian ores of Tealby and Claxby in Lincolnshire are similar. A coarser-grained iron-oolite (pisolitic iron-ore of some writers) occurs near Tremadoc, *etc.*, in North Wales, and is of Lower Palæozoic age.

If the grains of an oolitic ironstone be dissolved by acid, each leaves a shell or skeleton of silica soluble in caustic potash. This silica must have been introduced at some stage of the alteration of the original limestone. A similar siliceous skeleton is sometimes found in the grains of oolitic limestones where no ferruginous replacement has taken place, or, again, silica may more or less replace the calcareous matter between the grains[1]. Although *silicification* is perhaps less common than some of the other metasomatic changes noticed above, it is found in numerous limestones of various ages. Sometimes the replacement of carbonate of lime by silica is confined to the organic remains, but in other cases it affects the whole body of the rock (*e.g.* some cherts). Parts of the Carboniferous limestones of Clifton shew examples of oolitic grains and organic fragments replaced by a mixture of limonite and silica. Good examples of cherts formed by the silicification of limestone (matrix and fossils alike) are found in the Portland Beds of the South of England.

An almost purely siliceous rock from eastern Pennsylvania[2] shews a beautiful oolitic structure, each little sphere, about ·04 inch in diameter, consisting of numerous concentric coats surrounding a nucleus, and the interspaces being also occupied by silica. Here there seems to have been evidently a molecular replacement of carbonate of lime by silica, and indeed associated rocks shew various stages of partial replacement. Some cherts in the Durness Limestone of Sutherland tell the same story, the oolitic structure being still discernible (Stonechrubie near Inchnadamff). Similar oolitic cherts occur in the Corallian of Yorkshire.

Mr Rutley[3] believes silicification of dolomitic strata to have gone on very extensively in some cases, and explains thus the formation of certain novaculites (whetstones) and some other siliceous rocks.

Still another metasomatic change met with in some calcareous rocks is *phosphatization*. This usually affects some or all of the organic remains, or phosphatic nodules are formed having fossils of various kinds as nuclei. The phosphate of lime is presumably itself derived from organic bodies, but it is not clear to what extent it has been supplied contemporaneously with the deposit which contains the nodules. Deposits rich in phosphate occur at various horizons in the formations of this country: the Cambridge Greensand may be taken as an example,

[1] Chapman, *G. M.* 1893, 100–104.

[2] Barbour and Torrey, *Amer. Journ. Sci.* (1890) xl, 247–249, with figures.

[3] *Q. J. G. S.* (1894) l, 377–391, Pl. xix.

where the fossils are largely phosphatized and also serve to some extent as the nuclei of nodules. In other instances phosphate of lime occurs as casts of foraminifera[1] or as grains more or less definitely replacing those bodies[2]. Phosphatic deposits are now forming in the ocean, both within the littoral belt and in connection with the globigerina-ooze, *etc.*[3].

Some British limestones. After what has been said in the foregoing paragraphs a few remarks on some of the more important calcareous formations of this country will be sufficient to illustrate our subject.

The Bala Limestone of North Wales is sometimes a fine calcareous mud-stone, sometimes recrystallized. The most conspicuous organic fragments are those of crinoids, which are in places very abundant, and polyzoa are also found. The Hirnant Limestone[4] has a peculiar type of oolitic structure, the grains having a chalcedonic skeleton and concentric zones rendered opaque by finely divided carbon. The Coniston Limestone of Westmorland is in its purer parts usually recrystallized throughout to a granular mass in which the original characters are lost. In places it is dolomitized[5] (fig. 58). In its lower part it contains much non-calcareous material, chiefly volcanic, and at one horizon there is a breccia in which the enclosed fragments are of rhyolite, andesite, *etc.*, the matrix being calcareous[6]. At Keisley, in the Eden Valley district, the rock is in parts coarsely crystalline, matrix and fossils being recrystallized in common, so that, while the outlines of the larger fossils are preserved, all minute structures are destroyed[7]. At a lower horizon in the same district occur bands composed wholly of the little crustacean Beyrichia.

The Wenlock Limestone of Dudley, with a recrystallized matrix, still preserves abundant organic fragments, especially

1 Chapman, *Q. J. G. S.* (1892) xlviii, 514–518, Pl. xv.
2 Strahan, *Q. J. G. S.* (1891) xlvii, 357–362 (Chalk, Taplow).
3 Murray and Renard, Challenger Reports, *Deep-Sea Deposits* (1891), Pl. xx.
4 Fulcher, *G. M.*, 1892, 114–117, Pl. IV.
5 *Q. J. G. S.* (1893), xlix, 367.
6 *Ibid.* (1891) xlvii, 309, 310.
7 Nicholson and Lydekker, p. 20, fig. 5 *A*.

those of crinoids, entomostracans, trilobites, corals, polyzoans, and brachiopods. At Malvern the rock is largely oolitic, the grains being set in a recrystallized matrix, and sometimes themselves recrystallized (the Wych). Composite and broken oolitic grains also occur (Croft). The Aymestry Limestone, from Dr Sorby's description[1], is very like the Wenlock.

Dr Sorby[2] has pointed out many interesting features in the Devonian limestones of Devonshire. The recognizable organic fragments are chiefly of crinoids and corals, and the finely divided calcareous matter is probably derived from the degradation of coral skeletons. This fine material has often been recrystallized in the usual fashion, the impurities being segregated into patches of finer texture. Again, rhombohedral crystals of dolomite (often ferriferous) have frequently been formed in the rocks, and some have become true dolomite-rocks, while a little pyrites, partly oxidized, is not uncommon. Many of the rocks shew slaty cleavage in every respect similar to that noticed in argillaceous strata, and the deformation of the rock-masses is indicated by distortion of the crinoidal fragments, *etc.* Discontinuous slipping has also taken place on large and small scales, especially near Ilfracombe, and many remarkable structures have been set up[3]. A frequent feature is the occurrence of 'eye-shaped' or lenticular remnants of uncrushed crinoidal limestone in a limestone-slate (see below, fig. 72). Some rocks in this neighbourhood shew partial silicification[4].

The Carboniferous limestones of Clifton, Bristol, are, in different beds, composed largely of fragments of polyzoa and crinoids, tests of foraminifera, *etc.* Numerous oolitic beds occur, and in some of these Mr Wethered[5] has found the oolitic structure to be connected with the growth of Girvanella. In others the oolitic grains are in some measure replaced by iron-oxides and silica, and some of the organic fragments (especially of polyzoa) also shew a ferruginous replacement (fig. 55, *B*). The interstitial calcareous mud is usually recryst-

[1] *L.c.* p. 60.

[2] *Phil. Mag.* (1856) ser. 4, xi, 20–37. See also Wethered, *Q. J. G. S.* (1892) xlviii, 377–387, Pl. IX.

[3] Marr, *G. M.*, 1888, 218–221.

[4] Chapman, *G. M.*, 1893, 100–104.

[5] *Q. J. G. S.* (1890) xlvi, 270–274, Pl. XI.

allized as a rather coarse calcite-mosaic. In the Carboniferous limestones of the Forest of Dean Mr Wethered[1] finds remains of crinoids, polyzoa, and shells, valves of ostracods, spines of brachiopods, *etc.* Both here and in the Bristol district dolomitization occurs at some horizons.

The Mountain Limestone of the North of England is on the whole of similar character. The most frequent of the recognizable organic fragments are in many cases those of crinoids, and at some horizons in Derbyshire and Yorkshire these constitute the main bulk of the rock, but fragments of brachiopods, corals, polyzoa, and algæ also occur, and may be abundant[2], while foraminifera are often very plentiful[3]. Dr Sorby has pointed out that in Derbyshire some of the beds are pure dolomite-rocks[4]. The remains of shells, corals, and crinoids are dolomitized, as well as the matrix, their internal structure being quite obliterated and their outlines often obscured. In such a case the conversion of calcite to dolomite by secondary processes is evidently beyond doubt. Mr Rutley found that on treatment with dilute acid a rock from Matlock disintegrated, yielding perfect or corroded rhombs of dolomite.

Dolomite-rocks and dolomitic limestones occur at many localities in the Carboniferous of Ireland[5]. They are in general highly crystalline, and all trace of organic structures is obliterated. A common type seems to be that in which the predominant dolomite, in more or less imperfect crystals, is cemented by calcite. This becomes evident on weathering, when the removal of the calcite sets the dolomite crystals free. The rocks are always more or less cellular or porous, but the cavities are commonly filled, or lined in drusy fashion, by calcite. Similar phenomena occur in the Carboniferous limestones of the Isle of Man, and are beautifully exhibited on the

[1] *G. M.*, 1886, 529–540, Pl. XIV, XVI. For description of Carboniferous Limestones from N. Flintshire see Wethered in Morton's *Geology of Liverpool* (2nd ed. 1891), 25–27.

[2] Nicholson and Lydekker, p. 19, fig. 4.

[3] The Saccamina Limestone of Northumberland may be mentioned as a Carboniferous rock consisting essentially of foraminifera: *ibid.* p. 126, fig. 30. See also p. 21, fig. 6, an American foraminiferal limestone of Carboniferous age.

[4] See also Rutley, *Q. J. G. S.* (1894) l, 381, 382, Pl. XIX, figs. 5, 6.

[5] See Hardman, *Proc. Roy. Irish Acad.* (1876) ser. 2, ii, 723–726.

shore at Castletown and Poolvash. The resulting dolomite-rock is often quite coarsely crystalline.

As an example of a limestone-breccia or conglomerate may be cited the 'Brockram' which forms the base of the Permian in the Eden Valley. The fragments and pebbles of various sizes are derived from the Mountain Limestone, and many of them are dolomitized. These, with very numerous rolled quartz-grains, are enclosed in a matrix of crystalline calcite. The latter often shews in places little clear spherules giving a perfect black cross, but evidently of a different nature from the common type of oolitic grain. The Dolomitic Conglomerate of the Mendips belong to a higher horizon.

The Permian Magnesian Limestone is in general a true dolomite-rock, and in most cases all minute original structures have been lost in the changes which converted the rock to a granular mass of dolomite. When organic fragments are recognizable they are most frequently those of shells and polyzoa. Locally in South Yorkshire the latter bodies make up almost the whole of the rock (Brodsworth, Cadeby, *etc.*). Near Abergele in North Wales foraminifera and corals form a large part. Dr Sorby describes the Magnesian Limestone north of Nottingham as comparatively coarse-textured, with evident rhombohedral crystals. The usual type in Durham is often fine-grained, the elements being of irregular form. Sometimes an interlocking arrangement of the granules, aided by the presence of little vacant spaces, gives a certain flexibility to the rock[1] (Marsden). The little cavities or pores are, however, as in other dolomitic rocks, often occupied by crystalline calcite. The well-known nodules of Marsden and Sunderland, several inches in diameter and with well-marked radial crystallization, are of calcite with but little carbonate of magnesia[2]. The Magnesian Limestone is, as a rule, tolerably free from foreign detrital matter, but locally it becomes arenaceous. Dolomitic sandstones occur near Mansfield, and the attenuated representative of the Magnesian Limestone in Westmorland[3] is full of angular quartz-grains.

[1] Card, *G. M.*, 1892, 117–124.
[2] Garwood, *G. M.*, 1891, 434–440.
[3] Nicholson and Lydekker, p. 18, fig. 2 *B*.

In the Lower Oolites of the Cotteswold and Bath districts[1] fragments of shells, crinoids, and polyzoa, tests of foraminifera, and other organic remains are recognized in variable proportions. Most of these limestones are oolitic, but the original structure of the oolitic grains is often destroyed by recrystallization. In the best preserved examples Girvanella is detected at various horizons, and it is specially well exhibited in the coarse pisolite known as the 'Pea Grit'[2]. The rocks contain various small proportions of insoluble residue consisting of detrital mineral fragments (quartz, *etc.*).

The Lincolnshire Limestone and Millepore Oolite of the North of England[3] are made up largely of oolitic grains of the ordinary type, consisting of a nucleus of a shell-fragment, a quartz-grain, or a brown pellet of mud, surrounded by numerous iron-stained coats, in which a radial structure is sometimes discernible. The organic fragments include chips of brachiopods and Pecten, recrystallized fragments of aragonite shells, foraminifera, valves of ostracods, pieces of echinoderms, *etc.*, in different beds: *e.g.* abundant brachiopod spines in the Rhynchonella spinosa beds. The general matrix of fine calcareous mud is almost always converted into a crystalline calcite-mosaic with localisation of the ferruginous impurities, and most of the rocks contain a considerable amount of angular quartz-sand. This last feature is more prominent in the Scarborough Limestone and the Cornbrash. The former, especially in certain nodular bands, is often an ironstone consisting of minute rhombohedra of chalybite, with no calcite remaining except in the fragments of shells.

The Coral Oolite of Malton is another good specimen of an oolitic limestone with recrystallized matrix. Besides foraminifera, crinoid fragments, *etc.*, it contains abundant remains of aragonite gasteropods replaced by calcite mosaic. The oolitic grains are sometimes large enough to be termed pisolitic, but

[1] Wethered, *Q. J. G. S.* (1890) xlvi, 274–277, Pl. XI; (1891) xlvii, 550–569, Pl. XX.

[2] See also *G. M.*, 1889, 197, 198, Pl. VI.

[3] *Naturalist*, 1890, 300–304. For figures of various Lower Oolitic limestones see *Mem. Geol. Surv.*, *Jurassic Rocks of Britain*, vol. iv, Pl. II and explanation.

the Girvanella noticed by Mr Wethered[1] in the Osmington pisolite, near Weymouth, is not yet recorded from Yorkshire. The last-named author (*l.c.*) has described the Portland rocks with their recrystallized oolitic grains. The silicification of some beds in that district has already been referred to.

The microscopic characters of the English Chalk have been described by Dr Sorby[2], Messrs. W. Hill and Jukes-Browne[3], and others. The tests of foraminifera, and especially detached cells of Globigerina, are abundant in many examples[4], though they rarely form the chief constituent of the rock. The cells are empty in the soft chalk of the South, but filled with calcite in the hard chalk of Yorkshire. Molluscan fragments, and especially the detached shell-prisms of Inoceramus, are often well represented: in the Totternhoe Stone shell-fragments form 60 to 70 per cent. of the rock. In most cases, however, the great bulk of the rock consists of very finely divided calcareous material, the nature of which can be studied only by rubbing the chalk with water and examining the powder. Coccoliths abound in this fine mud[5], but the minute granules are mostly such as would come from the destruction and dissolution of aragonite shells, corals, *etc.* Foreign detrital matter is rare in the Chalk, except at certain horizons (*e.g.* the Hunstanton Rock). The Cambridge Greensand has rather large quartz-grains, with some mica. It also contains a considerable number of glauconite grains, usually as perfect internal casts of foraminifera[6], and glauconite occurs at some higher horizons in smaller quantity. Sponge spicules may be found in some examples. Those in the Lower Chalk of Berkshire and Wiltshire are sometimes preserved in the original colloid silica, sometimes replaced by calcite, while little globules of colloid silica (·0006 inch in diameter) occur in the rock.

[1] *G. M.*, 1889, 197, Pl. VI, fig. 9; *Q. J. G. S.* (1890) xlvi, 277–279; Pl. XI, figs. 6–8.

[2] *Q. J. G. S.* (1879) xxxv, Proc. 48, 49.

[3] *Q. J. G. S.* (1886–9) xlii, 228–230, Cambridge and Hertfordshire; 242, 243, Dover; xliii, 580–585, W. Suffolk and Norfolk; xliv, 355–357, Lincolnshire and Yorkshire; xlv, 406–413, Berkshire and Wiltshire.

[4] For figs. see Geikie, p. 122, fig. 22; Nicholson and Lydekker, p. 19, fig. 3.

[5] On coccoliths in the Chalk see Sorby, *Ann. Mag. Nat. Hist.* (1861) 3, viii, 193–200.

[6] Sollas, *Q. J. G. S.* (1872) xxviii, 399.

CHAPTER XIX.

PYROCLASTIC ROCKS.

THE fragmental volcanic rocks are in general the products of explosive action[1]. The ejected material varies from the finest dust to pieces several inches, or even feet, in diameter, but the coarsest types do not require special notice here.

What is known as *volcanic dust* or fine ash is no doubt partly due to the comminution of rock and crystals by friction during the explosion, but a great part of it must represent lava blown out from the vent in liquid form and solidified almost instantaneously in the air. It doubtless solidifies as glass, but may, of course, be subsequently devitrified. The bodies known as volcanic *bombs* and *lapilli* are of very various sizes. They may have spheroidal or more peculiar forms; or again they may be irregularly shaped or fitted together. Some kind of concentric structure, with a nucleus and an outer crust, is often seen, or the exterior may be scoriaceous. In many volcanic accumulations *crystals* play an important part. They are commonly idiomorphic, though frequently broken, and belong to the minerals common in lavas. They may sometimes be torn from solid rocks, but more generally they must have been contained in a fluid matrix before the eruption. We also find *rock fragments*, either angular or, in submarine deposits, partly rolled and worn. They are commonly of lava for the

[1] The exceptions ('flow-breccias,' *etc.*) are not important for our present purpose.

most part, shattered and blown out by the explosion, but we also find pieces of igneous rocks which must have come from greater depths, or fragments of slate, grit, limestone, *etc.*, representing strata broken through, and often shewing evident metamorphism. The larger 'ejected blocks' are frequently of these foreign or non-volcanic rocks.

The rocks formed by the accumulation of these various materials have received many names. The term *ash*, applied to the finer incoherent products of modern volcanoes, is sometimes used in a more extended sense; but the older, more or less compacted, deposits of ash-material are usually called *tuffs*. A large proportion of them were evidently laid down under water: subaërial accumulations have less frequently been preserved from destruction. Rosenbusch, in describing the ancient acid tuffs, divides them into compact tuffs, crystal-tuffs, and agglomeratic tuffs, and the division may be applied to rocks of other composition; but, since the relative proportions of dust, crystals, and lapilli, *etc.*, may vary to any extent, no precise divisional lines can be drawn. If angular rock-fragments be largely represented, the deposit is termed a *volcanic breccia*, or if the fragments be rounded, a *volcanic conglomerate*.

According to the nature of the material, the rocks may often be spoken of as 'rhyolite-tuff', 'trachyte-tuff', *etc.*, or, again, 'andesite-breccia', 'trachyte-conglomerate', and so forth; but, owing to the admixture of various materials, the rocks do not always correspond exactly even with contemporaneous lavas directly associated with them.

Further, when deposited under water, the volcanic material may become mixed with ordinary detritus or with calcareous matter, and so we have earthy tuffs, calcareous tuffs, *etc.*, some of which are fossiliferous.

General characters. Fragmental volcanic rocks have received much less minute study than lavas, and indeed present greater difficulty, requiring for the finer material the use of high magnifying powers.

Typical volcanic dust in a fresh state seems to consist essentially of glass-particles, with only a minor proportion of comminuted crystals and microlites. The glass-fragments have a peculiar structure and a characteristic form. This is due to the immense number of contained gas-bubbles, which are drawn out into minute tubes, causing the glass to break into prismatic shapes with a longitudinal striation. The glass is distinguished from comminuted felspar by the absence of true rectilineal boundaries and the isotropic character. The minute fragments are colourless, except in the case of basic glasses, which may be of a brown tint. According to Murray and Renard[1], the characteristic appearance of these glass fragments may be recognised even in excessively small particles (less than ·0002 inch), while the distinctive properties of most minerals cannot be detected in fragments of smaller dimensions than ·002 inch. The minerals commonly recognized are the familiar constituents of volcanic rocks—especially plagioclase, pyroxenes, and magnetite[2], for many of these very fine volcanic dusts are of the nature of pyroxene-andesite. The crystals are often coated with glass or have glass adherent to them.

FIG. 60. BASIC TUFF, ORDOVICIAN, WET SLEDDALE NEAR SHAP; ×20.

The bulk of the rock is of very fine particles, but encloses some rock-fragments and numerous crystals of felspar, which tend to stand perpendicularly to the lamination of the matrix [895].

[1] See especially *Nature* (1884), xxix, 585–589.
[2] Fouqué and Michel Lévy, *Min. Micr.*, Pl. XIII, fig. 4.

The authors named find precisely similar material to be widely distributed in modern deep-sea deposits, where it accumulates from the fall of wind-borne dust and the disintegration of floating pumice.

In tuffs formed not far from a volcanic centre, crystals of recognizable size, perfect or broken, are often embedded in a fine-textured matrix. These frequently shew a characteristic arrangement, standing with their long axes vertical or roughly perpendicular to the lamination of the matrix, as if dropped into their place from above (fig. 60).

In any except comparatively young tuffs the original character of the finely divided material is largely obscured by secondary changes, the loose texture of the deposits rendering them peculiarly liable to alteration. According to the nature of the rock, such minerals as quartz, sericitic mica, chlorite, calcite, *etc.*, are developed at the expense of the original dust. Silicification is very common in the acid tuffs. Fragments of lava naturally suffer less than the enclosing matrix, but if glassy they readily become altered. In particular the more basic glasses, such as basalt and augite-andesite, are hydrated and converted into the transparent brown or yellow, or more rarely green, substance known as palagonite[1].

In some cases it is very difficult to distinguish compact rhyolite-tuffs, silicified or otherwise altered, from rhyolites which have undergone similar changes, the lamination of the one and the flow-structure of the other often increasing the resemblance. When enclosed crystals occur, their characteristic orientation, as noted above, will often furnish a clue; or again the occurrence of fragments with concave outlines (Ger. Bogenstructur) is sufficiently suggestive. Old tuffs of andesitic and basic composition, when more or less cleaved and impregnated with secondary chlorite, calcite, and other substances, may sometimes be mistaken for crushed lavas of like composition, or *vice versâ*, unless distinct fragments, such as lapilli, can be detected. These lapilli can often be recognized by a

[1] For some discussion of the nature of this substance, see Zirkel, *Micro. Petr. Fortieth Parallel*, pp. 273–275 (1876). The basic glass which has not suffered hydration is sometimes termed sideromelane: see also Murray and Renard, *Chall. Rep.*, *Deep-Sea Deposits* (1891).

rounded outline, or a vesicular structure, or an opacity due to finely divided magnetite[1].

It will easily be understood that fine-textured tuffs may exhibit precisely the same phenomena of slaty cleavage as those seen in argillaceous sediments, while the coarser pyroclastic rocks (volcanic breccias and agglomerates) are more readily crushed than solid rocks such as lavas.

Some special examples. Without attempting to deal systematically with the great variety of tuffs, agglomerates, *etc.*, it will be sufficient to draw attention to a few, which have been already described, and illustrate various points of interest.

As typical of many fine *volcanic dusts*, we may take that which was spread over a vast extent of country after the great eruption of Krakatau in 1883. This has been described by several writers[2]. About nine-tenths of the material consists of glass fragments with the characteristic features noticed above. The remainder is of comminuted crystals of plagioclase, magnetite, enstatite, and augite, the whole having the composition of an acid pyroxene-andesite.

We pass on to notice a few consolidated deposits (tuffs) of various composition.

An interesting study of ancient *acid tuffs* has been made by Mügge in the Devonian of the Lenne district in Westphalia. The rocks are associated with old soda-rhyolites ('Keratophyre' of the author), and have a similar composition. They are for the most part of compact type, and, though considerably altered, still retain much that was characteristic in their original structure. In particular, they often shew very clearly the peculiar form of the constituent ash-particles, bounded by concave curves, which clearly suggest broken up pumice[3]. Crystal-tuffs are also found.

[1] *Cf.*, *e.g.*, Teall's figure of one of the Llanberis tuffs, Pl. XLV, fig. 1.

[2] Murray and Renard, *Nature* (1884) xxix, 585–589; Cole, *Proc. Geol. Ass.* (1884) viii, 332–335; Joly, *Proc. Roy. Dubl. Soc.* (1884) *N. S.* iv, 291–299, Pl. XII, XIII.

[3] *Neu. Jahrb. für Min.*, Beil. Bd. viii, Pl. XXIV, figs. 20, 21, *etc.* (1893). All the figures illustrating this paper are instructive.

Some of the Ordovician rhyolite-tuffs of Caernarvonshire have much resemblance to those just mentioned[1]. Others, there and in the Lake District, have evidently consisted of much more finely divided material, and have often lost all trace of their original characteristics by secondary changes. Embedded crystals usually occur (Glyder Fawr, *etc.*), but do not make up any large part of the mass. There are, however, beds made up very largely of small rock-fragments and broken crystals lying in a fine-textured matrix or united by a brown ferruginous paste. The rock-fragments are of various quartz-porphyries and granophyres, and sometimes detached spherules; the crystals are of acid felspar and decomposed augite (near Llanbedrog, *etc.*)[2]. Prof. Bonney[3] has described an agglomeratic type from the older rocks of the Llanberis district as consisting of fragments and lapilli of rhyolite and fragments of quartz and felspar embedded in an altered felspathic dust. Here some of the rock-fragments are of large size.

Among various other ancient rhyolite-tuffs in this country may be mentioned those of the Malverns (Knighton, *etc.*), in which an interesting feature is the development of veins and patches of clear secondary felspar, often shewing twin-lamellation. A similar thing is seen in some of the Lenne rocks mentioned above. In some examples from the Eden Valley (Wythwaite Top) the development of clear secondary felspar has proceeded further[4].

Some of the fine-textured rocks which have been styled 'porcellanite' and 'hälleflinta' are acid tuffs compacted by secondary silica and other substances. Examples occur in the St David's district (Clegyr Bridge, *etc.*) and in Charnwood Forest (Nanpanton). Rocks of the same general aspect in the Lake District (Bow Fell, *etc.*) are fine tuffs of intermediate composition.

Trachyte-tuffs of various types are known in many of the newer volcanic districts of Europe and America, but they have not often been minutely studied. The rock known as 'trass'

[1] *Neu. Jahrb. für Min.*, Beil. Bd. viii, Pl. xxvii, fig. 41.
[2] *Bala Volc. Ser. Caern.*, p. 27.
[3] *Q. J. G. S.* (1879) xxxv, 312.
[4] *Ibid.* (1891) xlvii, 515, 516.

is, at least in part, of this nature. In the Siebengebirge is a considerable development of trachyte-conglomerate. The leucitophyres of the Eifel district are accompanied by tuffs of corresponding nature. These contain fragments of leucitophyre, chiefly in the condition of pumice, and of the Devonian slates of the district, in a matrix of glass-particles with many crystals of leucite and crystals (often broken) of augite, sanidine, and other characteristic minerals of the lavas (Olbrück, Rieden, *etc.*).

A good example of a hornblendic *andesite-tuff* is extensively developed at Rhobell Fawr[1] in Merioneth, an old volcano probably of late Cambrian age. It is in great measure a crystal-tuff, the most conspicuous elements being perfect and broken crystals of brown hornblende and pale yellowish augite.

The majority of the Ordovician tuffs in the Lake District correspond in general composition with andesites and with basic andesites or basalts, but many of them have in addition angular fragments of rhyolite. Crystals of felspar are often seen, but do not make up a large part of the rocks, which are essentially of the compact type in most cases (fig. 60). Rolled pieces of lava of small dimensions may occur. In some localities the rocks consist mainly of a mixture of small lapilli with fragments of slate, grit, *etc.*, often metamorphosed. Mr Hutchings has described an example from Falcon Crag near Keswick[2]. The finer tuffs of the district are often cleaved and highly altered (see below).

The cleaved tuffs of Cader Idris[3] in Merioneth also contain plenty of slate-fragments with felspar crystals and particles of scoriaceous andesite-glass converted into green palagonite, all set in a fine ashy matrix.

Other Ordovician tuffs consist largely of little fragments of formerly glassy and sometimes pumiceous andesite, now converted into a palagonite-like material of yellow or brown colour (*e.g.*, Snead in Shropshire)[4].

[1] Cole, *G. M.*, 1893, 343.
[2] *G. M.*, 1891, 462.
[3] Cole and Jennings, *Q. J. G. S.* (1889) xlv, 423–431.
[4] Cole, *Q. J. G. S.* (1888) xliv, Pl. XI, fig. 5; *Aids to Pract. Geol.* p. 180, fig. 22.

Some interesting fragmental rocks of *basic* composition occur in the old volcanic series of St David's of early Cambrian or pre-Cambrian age. They are agglomeratic tuffs consisting chiefly of little fragments of basic lava, sometimes rounded but usually angular or subangular. In some there is very little matrix: it consists of fine *débris* of the same material as the larger fragments. Sir A. Geikie[1] has described specimens from Pen-y-foel and Pen-maen-melyn. Acid tuffs occur in the same series. These too are mostly of the agglomeratic type, and may be styled breccias, consisting largely of fragments of old rhyolitic lava (Clegyr Hill, *etc.*). More compact tuffs are also found.

Among the basaltic rocks crystal-tuffs seem to be almost unrepresented. A common type consists of lapilli of basalt (glassy or altered) cemented by calcite, aragonite, limonite, *etc.* Widely distributed is the *palagonite* type[2] of Waltershausen, described from Sicily, Iceland, the islands of the Pacific, *etc.* This consists chiefly of little fragments of altered glassy basalt, usually of brown colour, often vesicular, and sometimes enclosing a few crystals of augite, olivine, or basic plagioclase; while the cementing material is obtained from the decomposition of the fragments, or may include calcite derived from calcareous matter contemporaneously deposited or by infiltration from without. Such rocks are widely represented among the older formations in this country.

Submarine tuffs of intermediate and basic composition occur, for example, abundantly in the Carboniferous in the basin of the Firth of Forth. Most of them contain some admixture of detrital or calcareous matter, but characteristic examples of tuffs, and in particular of palagonite-tuffs, are found. As described by Sir A. Geikie[3], the bedded deposits consist of a fine-textured matrix enclosing fragments of lava. The latter are the *débris* of already consolidated

[1] *Q. J. G. S.* (1883) xxxix, 295–300, Pl. IX, figs. 1, 2.

[2] For figures of palagonite-tuffs see Zirkel, *Micro. Petr. Fortieth Parallel*, Pl. XII, figs. 3, 4. Rosenbusch, *Mass. Gest.*, Pl. VI, fig. 4.

[3] *Trans. Roy. Soc. Edin.* (1879) xxix, 513–516, Pl. XII, fig. 10. For examples of similar rocks of pre-Cambrian age, see R. D. Irving, *Copper-bearing Rocks of L. Superior*, Pl. XV.

rocks rather than typical lapilli: they are largely vesicular, not only at the margin but throughout, and the vesicles are often cut by the external surface of the fragment. Calcite, delessite, *etc.*, occupy the cavities. A common feature is fragments of a transparent green or yellowish material resembling serpentine, which is evidently an altered vesicular glass, and is referred to palagonite. The matrix of these rocks has probably consisted of finely divided material of the same general nature as the larger fragments, but its structure is completely obscured by secondary changes, and the mass is stained green or brown.

The tuffs associated with the Carboniferous olivine-dolerite lavas of Derbyshire are in great part composed of true lapilli, often bordered, and having numerous vesicles not broken by the outline of the lapillus[1]. The material is a brown glass with globulites and crystallites and with crystals of olivine or plagioclase. These minerals are often replaced by calcite, and the same substance fills the vesicles and forms the cement of the rock.

Tuffs, breccias, and agglomerates of Carboniferous age are well exhibited near Scarlet Point in the South of the Isle of Man[2]. Pumiceous and scoriaceous fragments of irregular form are crowded together with a calcareous cement, and there are also fragments of various sizes of a rock like the associated basalt.

Fine-grained tuffs, and in a less degree agglomerates, may receive, as already mentioned, a secondary *cleavage-structure* precisely similar to that observed in argillaceous rocks; and the cleavage is often accompanied by mineralogical changes. The cleaved tuffs or ash-slates of the Lake District have been noticed by Dr Sorby, and some of them described in detail by Mr Hutchings[3]. These rocks are of intermediate, and sometimes perhaps basic, composition, and the finely divided portions have undergone great secondary changes. Chlorite

[1] Arnold-Bemrose, *Q. J. G. S.* (1894) l, 625–642, Pl. xxiv, figs. 4, 5; xxv.
[2] Hobson, *Q. J. G. S.* (1891) xlvii, 442, 443.
[3] *G. M.*, 1892, 154–161, 218–228.

and dust or granules of calcite are often conspicuous, and when these have been removed by acid from the powdered rock, or from very thin slices, other minerals may be detected, especially minute sericitic mica, which gives bright polarization-tints. The needles of rutile, so characteristic of clay-slates, are not found, but there are sometimes granules of sphene (*e.g.* Kentmere). In some of these slates minute garnets play an important part (*e.g.* Mosedale, near Shap). In general there has been an abundant separation of silica, partly as quartz, partly perhaps as chalcedony.

This is the general character of the finest slates of the Lake District, which are evidently greatly altered from their original state. The coarser bands have a matrix of similar character enclosing lapilli and recognizable fragments of andesite and also of rhyolite. Some rocks of a comparatively coarse agglomeratic nature are worked for slates in Borrowdale.

The 'schaalsteins' of the Germans, as found in the Devonian of Nassau and the Harz, are in part cleaved basic tuffs, impregnated with calcite, chloritic products, *etc.*; but some of the rocks so named are apparently crushed lavas.

A remarkable *ultrabasic* breccia (Kimberley type) occurs as pipes, probably volcanic necks, in the diamond-fields of South Africa. Carvill Lewis[1] describes it as a breccia of porphyritic peridotite, consisting of phenocrysts of olivine, with some enstatite or bronzite, biotite, pyrope garnet, ilmenite, perofskite, *etc.*, in a serpentinous matrix. The rock is altered to a great depth into the 'blue ground' of the miners, in which the olivine is converted to serpentine, the mica to the so-called vaalite, *etc.* This is the matrix of the diamond.

Many tuffs have a calcareous cement. In some cases the calcite may have been derived from the destruction of lime-bearing silicates or introduced in solution from an extraneous source. There are, however, many submarine tuffs of all ages in which calcareous organisms have been accumulated contemporaneously with the volcanic material, giving rise to every gradation from a pure tuff to a pure limestone. Such deposits

[1] *G. M.*, 1887, 22–24; 1888, 129–131.

are forming at the present day in the neighbourhood of volcanic islands, and consolidated *calcareous tuffs*, often abounding in foraminifera, *etc.*, are beautifully represented among the Recent

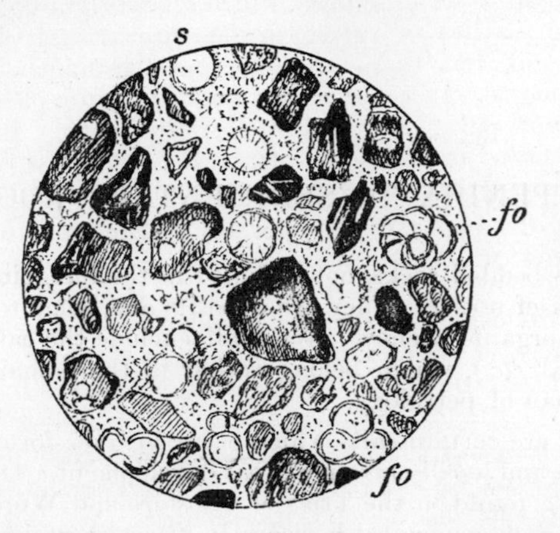

FIG. 61. CALCAREOUS TUFF, EUA, TONGA ISLANDS; ×20.

The fragments are mainly of brown-stained andesitic and basic lava, more or less glassy and altered to palagonite. These, with tests of foraminifera (*fo*), are enclosed in a calcareous matrix. Each foraminiferal chamber is occupied by calcite with radial fibrous structure, giving a perfect black cross between crossed nicols, and the same is seen in the little spherical bodies (*s*), which are doubtless detached chambers of *Globigerina* [1273].

strata of the Solomon Islands[1], the Tonga group[2] (fig. 61), Torres Straits, *etc.*

[1] Guppy, *Trans. Roy. Soc. Edin.* (1885) Pl. CXLV, *etc.*
[2] *G. M.*, 1891, 251–256.

APPENDIX TO SEDIMENTARY ROCKS.

A FEW bedded rocks, not included in the foregoing sections, deserve brief notice. They are deposits due, some to chemical, others to organic agency. We shall exclude the carbonaceous rocks (coal, *etc.*), which belong rather to the domain of fossil botany than of petrology.

There are certain salts which occur in beds, forming stratified rocks, and locally attain a great development. One of these is *rock-salt*, found in the Trias of Cheshire and Worcestershire and at various geological horizons in other countries. Besides admixture of other salts, the deposits contain more or less of clayey, organic, or other impurities. Rock-salt itself (sodium chloride) is colourless in slices, and has a strongly marked cubic cleavage and a low refractive index. It frequently contains microscopic brine-cavities of cubical shape.

Another mineral which may form a rock by itself is *gypsum*[1]. It occurs in allotriomorphic grains which may be very small. The strong clinopinacoidal cleavage is well marked; the refractive index is quite low, and the double refraction is weak (about equal to that of quartz). Gypsum is often associated with rock-salt.

The simple sulphate of lime *anhydrite* is also found as a rock (Val Canaria in Switzerland, *etc.*), building allotriomorphic to partly idiomorphic crystals or fibrous aggregates. The two strong pinacoidal cleavages are well marked; the refractive

[1] Hammerschmidt notes that the heating of the Canada balsam in mounting may cause partial dehydration of the mineral, giving rise to little matted aggregates of anhydrite.

index is low, but the double refraction very strong (equal to that of muscovite). The rock often encloses grains of rock-salt or of dolomite and other minerals. It is specially liable to conversion into gypsum, which may be seen in various stages, veins and patches of the latter mineral traversing the anhydrite mass. This involves an increase of bulk and phenomena of disruption.

Another mineral which sometimes forms a simple rock is *chalybite* or siderite, the ferrous carbonate. We have already seen that some iron-stones of this composition have been formed by metasomatic processes from limestones, but in other cases, such as the ironstone bands in the Coal-measures of Yorkshire, *etc.*, there is no evidence of such an origin. The mineral may be mixed with other carbonates in smaller proportions and with a variable quantity of argillaceous matter (clay-ironstone). Chalybite has the rhombohedral cleavage of the calcite-group of carbonates, and in its brownish yellow colour resembles dolomite.

Some siliceous rocks. Here we may notice certain siliceous rocks which do not, at least in the main, result from pseudomorphism of limestones. Some well-known examples of *cherts* fall under this head, the silica being derived from siliceous sponges, recognizable remains of which still form an important part of the rock. In the chert-beds of the Upper Greensand[1] in the Isle of Wight the sponge-spicules sometimes remain in their original condition, consisting of colloid (isotropic) silica, but more usually spicules and matrix are alike converted to chalcedony. The sponge-beds of similar age in the Weald district consist largely of colloid silica, but the spicules are represented by empty casts. The cherts of the Carboniferous limestones of Ireland[2] are found, in the best-preserved specimens, to consist largely of sponge-spicules, the matrix being also siliceous and doubtless derived from the dissolution of other spicules. Here the silica is always in the condition of

[1] Hinde, *Phil. Trans. Roy. Soc.* (1885), pp. 447, 448, Pl. 40; for abstract see *Q. J. G. S.* (1889) xlv, 406, 407. Nicholson and Lydekker, p. 159, fig. 51.

[2] Hinde, *G. M.*, 1887, 441–443.

chalcedony or quartz. The Yoredale cherts of Yorkshire and North-Wales are of similar character, with better preserved sponge-remains, and the same seems to be true of the Carboniferous cherts (Fr. phthanites) of Belgium, and of some other countries, including some of the 'Kieselschiefer' of the Germans.

Of great interest are the deep-sea deposits which give rise to siliceous rocks. The 'Challenger' Expedition[1] has shewn that these occur over extensive tracts of the ocean-floor in its deepest portions. Characteristic types are the *diatom-ooze*, essentially an accumulation of the frustules of diatoms[2], and the *radiolarian ooze*, made up mainly of the tests of radiolaria. There may be some admixture of finely divided volcanic material or its decomposition-products or of foraminiferal remains. The equivalents of this radiolarian ooze are found in Recent and Tertiary *radiolarian earths*, such as those of Barbados[3] and Trinidad[4], and, in a compacted form, in the *radiolarian cherts* of some of the older formations. The Ordovician cherts of the South of Scotland, described by Dr Hinde[5], shew in slices a faint cloudy appearance, giving a mottled effect between crossed nicols, but are frequently veined and stained with dark brown. In the transparent parts the radiolaria shew as shadowy circles defined by their interior being somewhat lighter than the surrounding matrix. In the stained parts the tests are replaced by a dark substance, and may retain much of their original structure.

A peculiar type of siliceous deposit is the *sinter* of the hot

[1] See especially Murray and Renard, *Chall. Rep.*, *Deep-Sea Deposits* (1891) with plates (Pl. xv. *etc.*).

[2] Diatomaceous deposits of more limited extent occur under different circumstances, especially in connection with peat at numerous Scottish localities; see Macadam, *M. M.* (1884) vi, 87–89; also in the Tertiary of Virginia, *etc.* The so-called Tripoli-earth is of this nature, and of lacustrine origin.

[3] See Jukes-Browne and Harrison, *Q. J. G. S.* (1892) xlviii, 174, 175; Nicholson and Lydekker, p. 34, fig. 12.

[4] Gregory, *Q. J. G. S.* (1892) xlviii, 538, 539. On a radiolarian earth from S. Australia see Hinde, *Q. J. G. S.* (1893) xlix, 221, Pl. v.

[5] *Ann. Mag. Nat. Hist.* (1890) ser. 6, vi, 41–47, Pl. III, IV. On a somewhat similar rock from Mullion Island, Cornwall, see Hinde, *Q. J. G. S.* (1893) xlix, 215, Pl. IV.

springs and geysers of the Yellowstone Park, Iceland, and New Zealand. Mr Weed[1] has shewn that this material, consisting of colloid silica, is in great part secreted by filamentous algæ (Leptothrix, *etc.*). The resulting sinter or geyserite does not always shew clear organic structures. Sinter is formed also in the same places by the evaporation of the water in which the silica was carried in alkaline solution.

[1] *Amer. Journ. Sci.* (1889) xxxvii, 351; more fully in 9*th Ann. Rep. U. S. Geol. Surv.* pp. 613–676.

E. METAMORPHISM.

Using the term 'metamorphism' in a broad sense, we understand by it the production of new minerals, or new structures, or both, in pre-existing rock-masses. We must limit such a conception by supposing on the one hand that the changes produced are sufficient to give a distinctive new character to the rock as a whole, and on the other hand that they do not involve the loss of individuality of a rock-mass (*e.g.* bodily fusion must be excluded).

It is customary to distinguish *thermal* metamorphism, due to heat, and *dynamic* metamorphism, due to pressure. These can to some extent be considered separately, and we shall examine some of their results in the following pages. But, before doing so, we must notice that very important changes, which cannot reasonably be excluded from the domain of metamorphism, are set up in rock-masses without the intervention of either high temperature or great mechanical force. Many of these changes depend upon the access of circulating waters in communication with the atmosphere, and we may, if we please, roughly group them as meteoric or *atmospheric* metamorphism. In most cases, however, these processes involve some change in the total composition of the rocks affected, either a loss of some constituents or an addition of others (water, oxygen, carbonic acid, and other substances): in other words there is often *metasomatism* as well as metamorphism.

The common weathering-products of igneous rocks are results of such processes, but it is convenient, as already remarked, to restrict the term metamorphism to cases in which the general mass of a rock is considerably altered: the serpentine-rocks are an example. It is important to observe, however, that the minerals produced by secondary actions of the kind here contemplated include some which are also common as original constituents of igneous rocks: we have already mentioned the occurrence in this way of secondary quartz, felspars, hornblende, *etc.* There is a frequent tendency of the new-formed substance to form as a crystalline extension of pre-existing crystals or grains of the same mineral (like the quartz in many quartzites); or again for a pre-existing mineral to be extended by an outgrowth of some allied mineral with the same crystalline orientation: *e.g.* one kind of plagioclase felspar may receive an extension of another kind, augite of hornblende, allanite of epidote.

The most striking examples of what we have termed atmospheric metamorphism and metasomatism are found among the sedimentary rocks. We have already remarked the conversion of sandstones to quartzites, the recrystallization of limestones[1] and their replacement by dolomite, iron-compounds, silica, *etc.*, and we have seen that very many argillaceous sediments have undergone extensive or almost complete reconstitution since they became strata.

More important is the evidence of the formation of crystalline schists on an extensive scale by metasomatic changes alone, described by Prof. Van Hise in the Lake Superior region. In the upper part of the Penokee Iron-bearing Series[2] of Michigan and Wisconsin felspathic grits, greywackes, *etc.*, are traced into finely crystalline mica-schists, with biotite and muscovite, all relics of the clastic structure being finally obliterated. In the lower members of the same series[3] rocks consisting of impure

[1] Stefani attributes to the influence of circulating waters the formation from Triassic limestones of the famous Carrara marble in the Apuan Alps: see *G. M.*, 1890, 372, 373 (*Abstr.*).

[2] Van Hise, *Amer. Journ. Sci.* (1886) xxxi, 453–459; Irving and Van Hise, *Penokee Iron-bearing Series* in 10*th Ann. Rep. U.S. Geol. Surv.* (1890) 423–435, Pl. XXXVIII–XLII. For an abstract, see *Etudes sur les schistes cristallins* in *Rep. Congr. Géol. Internat.* (*Lond.*, 1891).

[3] Van Hise, *Amer. Journ. Sci.* (1889) xxxvii, 32–47; Irving and Van Hise, *l.c.* *Cf.* Hudleston, *Proc. Geol. Assoc.* (1889) xi, 133–138.

carbonates mixed with chert have been converted into ferruginous quartz-schists, magnetite- and hæmatite-schists, magnetite- and hæmatite-bearing actinolite-schists, *etc.*, also by metasomatic processes (silicification and other replacements), apparently without the conditions of either thermal or dynamic metamorphism.

We may now pass on to such changes affecting rock-masses as are more usually understood by the term metamorphism as employed in text-books. These changes are in part mineralogical (in most cases without any very important metasomatism), in part structural. These two lines of change are so connected that they cannot be considered quite separately: roughly we may say that mineralogical modifications are the more prominent in thermal metamorphism, and structural in dynamic.

While treating in turn the chief features of thermal and of dynamic metamorphism, we must remember that their effects may be associated or superposed in the same area, and the assigning of particular mineralogical changes to one or the other cause is in many cases still a question for discussion.

CHAPTER XX.

THERMAL METAMORPHISM.

UNDER this head we include all changes produced in pre-existing rock-masses by the influence of high temperature[1]. In the simplest case this is brought about by the intrusion of an igneous magma in the neighbourhood ('contact' or 'local' metamorphism of many authors); but we must also include the effects of heat mechanically generated (thermal being then associated with dynamic phenomena), and those due to the internal heat of the Earth in a rise of the isogeotherms. These latter especially may affect rock-masses on a regional scale. We shall here avoid complication by drawing our examples, so far as is possible, from cases of thermal metamorphism produced by igneous intrusions.

Characteristic minerals. It will be convenient to refer briefly to the commoner minerals formed in thermal metamorphism, some of them being unknown or rare in igneous rocks. *Quartz* and felspars are widely distributed in metamorphic rocks of various kinds. The felspars include *orthoclase*, *albite*, *anorthite*, and probably intermediate members of the plagioclase series. They are often perfectly clear, and when they occur as minute shapeless granules in a mosaic they may easily be mistaken for quartz without special optical tests. The larger grains shew the cleavage and sometimes characteristic twinning or some approach to crystal outline.

[1] For a discussion of various questions concerning thermal metamorphism see *Science Progress* (1894) ii, 185–201, 290–303.

Both *muscovite* and *biotite* are found in metamorphosed rocks, the latter being very widely distributed. It is apparently a haughtonite and always strongly pleochroic, with a deep reddish-brown colour or, for vibrations parallel to the cleavage-traces, a very deep brown with a noticeable greenish tone. Intensely pleochroic haloes surround certain inclusions. Less usual than brown mica is a green *ripidolite* or a yellowish or greenish variety of *chlorite*. Exceptionally we find the manganese-bearing chloritoid mineral *ottrelite*[1]: it builds flakes without special orientation, and freely encloses impurities: the lamellar twinning parallel to the base and a modification of hour-glass structure are noticeable[2].

Highly characteristic of the metamorphism of argillaceous and some other rocks are silicates rich in alumina. *Andalusite* forms more or less idiomorphic crystals with the prism-form and usually some traces of the prismatic cleavage. It is recognized by its moderately high refractive index with low double refraction (about the same as in labradorite) and straight extinction. When it shews any colour, it is pleochroic, giving a rose tint for longitudinal and a very faint green for transverse vibrations. It may be quite clear, or may contain numerous inclusions, certain enclosed minerals being surrounded by a pleochroic halo (bright yellow to colourless). In *chiastolite*[3] the elongated crystals contain a large amount of foreign matter, apparently carbonaceous, arranged in the fashion peculiar to the mineral (fig. 64). *Sillimanite* builds elongated prisms or needles, which in shape, cross-fracture, and refractive index resemble apatite, but have a much stronger birefringence (fig. 62). They are often crowded together in matted aggregates imbedded in quartz ('Faserkiesel' or 'quartz sillimanitisé')[4]. *Cyanite*[5] or disthene is found less commonly, building more or less rounded crystals or grains, with pinacoidal cleavage

[1] See Gosselet, *M. M.* (1889) viii, 210 (*Abstr.*); Hutchings, *G. M.*, 1889, 214; Whittle, *Amer. Journ. Sci.* (1892) xliv, 270–277.

[2] Rosenbusch-Iddings, Pl. XXII, figs. 5, 6.

[3] *Ibid.*, Pl. XVII, fig. 3; Teall, Pl. XXXIII, fig. 2; Fouqué and Lévy, Pl. III, fig. 1.

[4] Barrow, *Q. J. G. S.* (1893) xlix, 338, Pl. XVI, figs. 1, 2; Zirkel, *Micro. Petr. Fortieth Parallel*, Pl. II, fig. 1; Rosenbusch-Iddings, Pl. XVII, fig. 4.

[5] Barrow, *l. c.*, 338, 339, Pl. XVI, figs. 3, 4.

and a cross-fracture corresponding to a gliding-plane. It is colourless or pale blue with pleochroism in thin sections, and, owing to its high refractive index, shews a strong relief. Longitudinal sections give extinction-angles up to 31°. *Staurolite* forms good crystals, the larger ones always crowded with various inclusions[1]. When fresh, it is yellowish or reddish-brown with distinct pleochroism and strong refringence and birefringence. This mineral, however, and in varying degree all the aluminous silicates, are very liable to decomposition, the characteristic product being white mica in minute scales (the 'shimmer-aggregate' of Barrow[2]). *Cordierite* is often not easily recognized. It builds pseudo-hexagonal prisms or, more commonly, shapeless grains, and basal sections only sometimes shew the curious triple twinning[3]. The mineral rarely shews its colour and pleochroism in thin slices, but is sometimes stained of a yellow tint. The refractive index and double refraction are low.

The metamorphism of calcareous rocks gives rise to numerous silicates rich in lime, or in lime and magnesia. The pure lime-silicate *wollastonite* is colourless in thin slices, and shews lower refringence and birefringence than the augites. It is further distinguished by having its two principal cleavages and its direction of elongation perpendicular to its plane of symmetry, and consequently giving straight extinction. As a rule, it occurs in quite small imperfect crystals. The augites of metamorphosed limestones, *etc.*, are either non-aluminous (*salite*) or aluminous (*omphacite*). They build imperfect crystals or crystalline patches, take part in a finely granular mosaic, or occur as little globules enclosed in other minerals. The crystals are occasionally twinned on the usual law. The green colour is often imperceptible in thin slices. Both salite and omphacite give extinction-angles of 38° or 40°, and it is not always possible to discriminate them, though the former is sometimes betrayed by its partial conversion into serpentine. The most common amphibole in these rocks is a colourless *tremolite* in imperfect crystals, crystalline patches, veins, or

[1] Zirkel, *Micro. Petr. Fortieth Parallel*, Pl. II, fig. 3. On arrangement of inclusions, see Penfield and Pratt, *Amer. Journ. Sci.* (1894) xlvii, 81–89.

[2] *Q. J. G. S.* (1893) xlix, 340, Pl. XVI, fig. 5.

[3] Rosenbusch-Iddings, Pl. XIX, fig. 3.

sheaf-like groupings. It may shew a fibrous structure or a good hornblende-cleavage, and a rough cross-fracture is also common. Green *hornblende* and blade-like *actinolite* are found in some rocks. The lime-garnet *grossularite* forms well-bounded crystals, often of considerable size, with included pyroxene granules, *etc.* It is often birefringent, and further shews between crossed nicols a polysynthetic twinning of a remarkable kind[1]. With this structure goes a strongly marked zonary banding, the concentric zones differing in birefringence. *Idocrase* occurs either in well-built crystals or in shapeless plates enclosing other minerals. The cleavage and colour are usually not to be observed in thin sections. The birefringence is variable, and a crystal often shews bands or lamellæ differing in interference-colours. *Zoisite* occurs in little prisms often grouped in sheaf-like fashion. It is characterized by longitudinal cleavage-traces, high refractive index, low polarization-tints, and straight extinction. *Epidote*, often associated with the last mineral, is usually in shapeless grains or granular aggregates, though it may present crystal-boundaries towards calcite, *etc.* The cleavages are well-marked, the two sets of traces intersecting at about 65° in a cross-section. Twinning is uncommon. The larger crystals shew the yellow colour and pleochroism. Other distinctive characters are the high refractive index, very brilliant polarization-tints, and straight extinction in longitudinal sections.

Among the other products of thermal metamorphism may be mentioned common garnet, chlorite, dipyre, magnetite and ilmenite, pyrite and pyrrhotite, sphene, rutile and anatase, spinels, corundum, and graphite. Further, the formation of a certain amount of isotropic matter is characteristic in some cases[2].

As a special mineral formed in metamorphosed rocks near an igneous intrusion may be noticed *tourmaline*. This mineral occurs in little grains, often in veins which represent cracks, or sometimes very abundantly as a constituent of a kind of

[1] See Rosenbusch-Iddings, Pl. XIV, fig. 2; also, for numerous figures, Klein, *Neu. Jahrb.*, 1883, i, Pl. VII–IX.

[2] On this and some other points see Hutchings, *G. M.*, 1894, 36–45, 64–75.

contact-breccia. It is restricted to the neighbourhood of acid intrusions, and depends on an actual introduction of certain materials from the igneous magma. White mica has sometimes a similar occurrence.

Metamorphism of arenaceous rocks. The effects of thermal metamorphism in arenaceous rocks are simple or complex according to the homogeneous or heterogeneous nature of the deposits affected. In a pure quartz-sandstone or grit there are no degrees of metamorphism possible. If the temperature be sufficiently high, the whole will be recrystallized into a clear quartz-mosaic without a trace of the original clastic character. Short of this change, the sandstone will be unaltered, except in such minor points as the expulsion of the water from the fluid-pores of the quartz, an effect noticed by Sorby at Salisbury Crags. The homogeneous quartzite resulting from the complete metamorphism of a pure quartzose rock is not difficult to distinguish from a quartzite formed by the deposition of interstitial quartz. There is no distinction of original grains and cementing material, no secondary growth upon original nuclei, but each element of the mosaic is clear and homogeneous, presenting an irregular boundary which fits into the inequalities of the adjoining elements. Such quartzites are locally produced in the Skiddaw grits abutting on the large granophyre mass at Ennerdale, in the Carboniferous sandstones near the Whin Sill of Teesdale, and in many other localities.

If the original sediment contained felspar grains, not much altered, as well as quartz, the felspar is recrystallized with the quartz, and is liable to be overlooked in the resulting mosaic without careful examination.

Where a quartzose sandstone or grit has contained scattered decomposition-products, such as kaolin, calcite, and chloritoid minerals, in small quantity, metamorphism produces a quartzite with granules of some accessory mineral. Thus, near the Shap granite, the grits in the Coniston Flag group have been transformed into a quartzite with granules of colourless pyroxene, formed from kaolin and calcite. Similarly the chloritoid minerals give rise to brown mica. A curious green mica occurs in the quartzite of Clova in Forfarshire.

The metamorphism of a specially pure type of siliceous rock has been described by Mr Horne[1] in the case of the Arenig radiolarian cherts of the south of Scotland, as they approach the Loch Doon granite. The final result is a mosaic of granular quartz with numerous minute round inclusions of biotite.

If the original rock was more impure, containing plenty of aluminous and other substances, the product of metamorphism ceases to have any apparent resemblance to a quartzite. Silicates of alumina, garnet, micas, *etc.*, may be extensively produced, and the metamorphosed rock assume the aspect of a

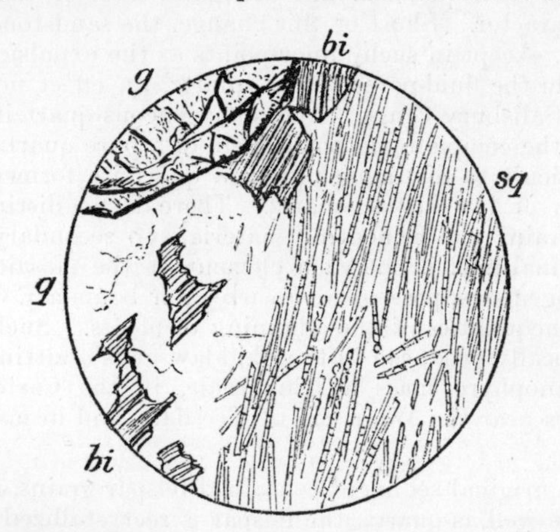

FIG. 62. GARNET-SILLIMANITE-SCHIST OR GNEISS, A HIGHLY METAMORPHOSED GRIT, CLOVA, FORFARSHIRE; × 20.

The right half of the figure shews an area of clear quartz full of little prisms of sillimanite with characteristic cross-fracture (*sq*): to the left are clear quartz (*q*), biotite (*bi*), and part of a large garnet (*g*) [1808].

fine or even a coarse gneiss (fig. 62). Remarkable examples are presented by the Silurian grits and flags around the Old Red Sandstone granite of New Galloway[2]. Here the chief constituents are quartz, muscovite, a deep brown biotite, and

[1] *Rep. Brit. Assoc.* for 1892, 712.

[2] Miss M. I. Gardiner, *Q. J. G. S.* (1890) xlvi, 569–580.

red garnet (colourless in slices), felspar being only subordinate. The garnets, except at the margin of each crystal, are crowded with minute granular inclusions: they tend to occur in clusters moulded by clear quartz, a frequent association in many metamorphic rocks. Nearer to the granite the texture of the rock becomes coarser, and the muscovite and quartz are seen to be crowded with narrow needles of sillimanite up to ·01 inch long. The same minerals as before are present, with a few crystals

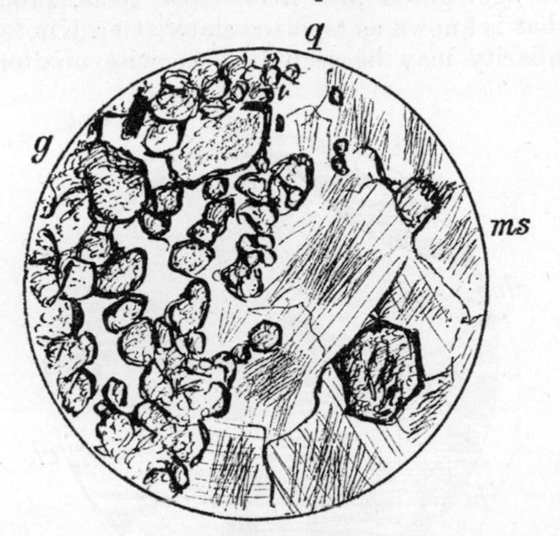

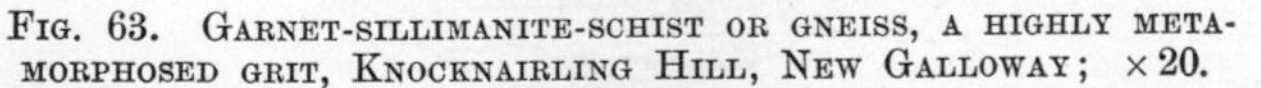
FIG. 63. GARNET-SILLIMANITE-SCHIST OR GNEISS, A HIGHLY METAMORPHOSED GRIT, KNOCKNAIRLING HILL, NEW GALLOWAY; ×20.

The figure shews portions of two lenticular streaks, one consisting essentially of muscovite crowded with minute needles of sillimanite (*ms*), the other of garnet (*g*) set in clear quartz (*q*) [1173].

of plagioclase and rarely a little brown tourmaline. At a hundred yards from the granite margin the texture is very coarse, the abundant white mica building plates half an inch in length and relatively thick. Dense matted aggregates of sillimanite needles occupy the interior of the quartz and muscovite, leaving the borders of the crystals clear. Some of the most altered rocks shew bands or streaks rich in particular minerals, such as lenticular patches of garnet set in clear quartz or streaks composed essentially of muscovite and sillimanite, dark mica being less plentiful (fig. 63).

Metamorphism of argillaceous rocks. The effects of thermal metamorphism in clays, shales, or slates depend in the early stages of alteration on the mineralogical, and in the later stages on the chemical, composition of the rocks affected.

In strata containing carbonaceous matter, this is one of the first ingredients to suffer change. It is either dissipated and expelled or converted into graphite. The latter is in some cases aggregated into little dark spots, producing one type of what is known as 'spotted slate' (Ger. Knotenschiefer). This peculiarity may be seen in otherwise unaltered strata,

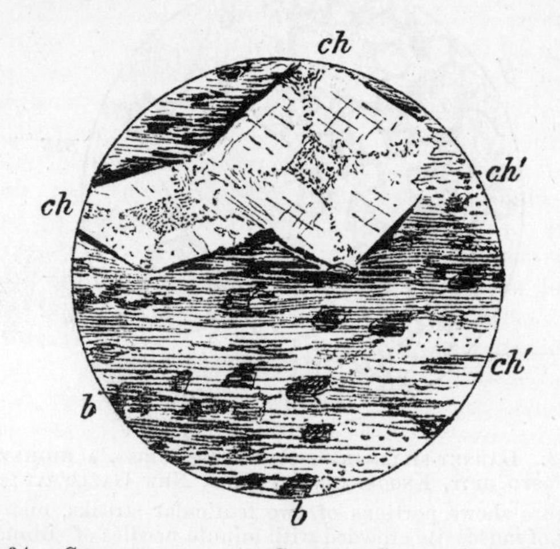

FIG. 64. CHIASTOLITE-SLATE, SKIDDAW SLATE METAMORPHOSED BY GRANITE, BANNERDALE, SKIDDAW; ×20.

Besides the good cross-sections of chiastolite (*ch*), shewing characteristic arrangement of enclosed impurities, there are imperfectly developed crystals (*ch'*) clearly detected by using polarized light. In the general mass of the rock the chief metamorphic effect is the production of little flakes of biotite (*b*) [1111].

and it disappears with advancing metamorphism. The minute needles of rutile so abundant in slates also seem to be rather readily affected, giving place to stouter crystals of the same mineral, or less commonly to anatase or brookite. Another early effect of metamorphism is the production of little flakes of brown mica (probably the haughtonite variety of biotite) from chloritoid substances, *etc.* With this there may be a crystallization of iron-ores (magnetite or pyrites). In some cases a chlorite mineral or ottrelite is formed instead of the mica. In rocks rich in alumina chiastolite is produced concurrently with biotite[1], *e.g.* in the Skiddaw district (Bannerdale, Roughton Gill, *etc.*, fig. 64).

With advancing metamorphism graphitic spots and chiastolite-crystals are lost, and the metamorphism begins to affect the whole body of the rock, the chief products formed being usually quartz and biotite. Of these the latter often has its flakes oriented in accordance with the original lamination or cleavage of the rock, and we have thus one type of mica-schist (Ger. Glimmerschiefer). These rocks may have no trace of the original clastic nature of the deposit, except perhaps some minute angular quartz-grains. They sometimes shew a spotted character quite different from that mentioned above, and consisting in little ovoid spaces free, or relatively free, from the flakes of biotite which crowd the rest of the rock. Such spaces often shew distinctly crystalline properties, giving extinction parallel with their length, and in many cases, at least, they are ill-developed crystals of andalusite. They may be observed in the Skiddaws of the Caldew and Glendcraterra valleys. When andalusite is better developed, it appears in clear crystal-grains or in crystal-plates enclosing other minerals: both forms are seen in the Skiddaw district, where andalusite-mica-schists have been extensively formed[2]. Other minerals, such as white mica and little garnets, occur more locally (Sinen Gill, Grainsgill, *etc.*).

As another example of well-marked spots due to the development of imperfect crystals we may take the Coniston Flags near their contact with the Shap granite[3]. Here the spots are small and ovoid, with numerous inclusions, but give a distinctly crystalline reaction, the essential mineral ex-

[1] For good coloured figures see Teall, Pl. XXXIII, fig. 2; Fouqué and Lévy, Pl. III, fig. 1.

[2] On the various stages of metamorphism in the Skiddaw district, see especially Rosenbusch (translated) in 'Naturalist', 1892, 119, 120.

[3] *Q. J. G. S.* (1891) xlvii, 320, Pl. XII, fig. 5.

tinguishing parallel to the length of the spots. Mr Hutchings finds it to be cordierite[1]. The same mineral forms somewhat larger spots in some of the metamorphosed Skiddaw Slates of the Caldew valley, and here some of the imperfect crystals shew the characteristic composite twinning (Swineside)[2].

Various types of spotted and flecked rocks due to metamorphism have been styled spilosite, Fleckschiefer, Fruchtschiefer, Garbenschiefer, *etc.*, and shew spots and patches of

FIG. 65. ANDALUSITE-MICA-SCHIST, METAMORPHOSED SKIDDAW SLATE, CLOSE TO GRANITE, SINEN GILL, SKIDDAW; ×20.

The rock consists of andalusite, biotite, and quartz, with subordinate muscovite and magnetite. It has not a very marked schistose character, and would be styled Hornfels by the German writers. All the lower half of the figure is occupied by a large crystal-plate of andalusite, enclosing numerous flakes of mica and needles of sillimanite [1446].

very various dimensions. In some they are evidently ill-formed crystals (*e.g.*, cordierite, Tirpersdorf, Saxony); in others the true nature of the spots is not very clearly understood. Mr Teall[3] compares with the typical 'spilosite' of the Harz

[1] Hutchings, *G. M.*, 1894, 65.
[2] *G. M.*, 1894, 169.
[3] *Brit. Petr.*, 218.

some slates near Tremadoc altered by large sheets of diabase. Here the spots are almost invisible in a slice viewed in ordinary light, but become conspicuously dark between crossed nicols. This seems to be due to numerous minute overlapping scales of chlorite[1]. A micaceous mineral occurs more sparingly, and an aggregate of granules having the refraction and double refraction of quartz and felspar. Similar phenomena are seen in other parts of North Wales, *e.g.*, near the granite of Ffestiniog.

In extreme cases of metamorphism the rocks lose all spotted, and frequently all banded and schistose, structures, passing sometimes into an extremely compact, fine-textured mass of quartz, micas, iron-ores, *etc.* (Ger. Hornfels, Fr.

FIG. 66. GRAPHITIC MICA-SCHIST, BLAIR ATHOLE, PERTHSHIRE; CUT PERPENDICULARLY TO THE SCHISTOSITY; × 20.

The rock consists mainly of quartz and sericitic mica, with some finely divided graphite. There are also numerous dodecahedra of garnet, each in the centre of a lenticular streak or 'eye' of quartz [1834].

cornéenne, hornstone of some writers). Andalusite, garnet, *etc.*, characterize different types (Ger. Andalusithornfels, Granathornfels, *etc.*). Some highly metamorphosed strata,

[1] Lossen believed that the spilosite of the Harz has a base of some isotropic substance, but this seems doubtful.

however, have a marked schistose character, usually due to micas of sericitic habit following old structural planes in the rock. Dark mica usually predominates, but white is also frequent. Red garnet is common in mica-schists of this kind, and other minerals may occur, according to the original chemical composition of the rock. A well-marked zone of graphitic mica-schists is known in the Central Highlands, and shews the characters of a thermally metamorphosed rock (fig. 66). The graphite doubtless represents carbonaceous matter of organic origin.

In slates which contained a considerable amount of finely divided felspathic matter, or at least had not become much impoverished in alkalies, the phenomena of metamorphism are somewhat different from those sketched above. Chiastolite is not formed, and andalusite does not usually figure largely in the more metamorphosed rocks, while white mica occurs abundantly with the biotite or to its exclusion. A good example of the type characterized by white mica is afforded by the slates of Charnwood Forest near a granitic intrusion at Brazil Wood[1]. Here the ragged flakes of muscovite enclose subordinate biotite with parallel intergrowth: a chlorite is also present, besides clear quartz and granules of opaque iron-ore.

Any peculiarity in the composition of the original strata gives rise in general to appropriate minerals during metamorphism. Thus shales or slates having any notable proportion of calcareous matter generate lime-bearing silicates, such as tremolite, *etc.* (see below, p. 271). But in many cases of contact-metamorphism material introduced into the metamorphosed rocks from an invading magma has given origin to special minerals not dependent on the nature of the strata affected. The commonest of these special minerals is tourmaline. It has been formed abundantly in many of the slates bordering the granitic intrusions of Cornwall[2] and Devon.

[1] Bonney, *Q. J. G. S.* (1877) xxxiii, 783. These slates are probably composed in great part of volcanic material.

[2] Allport, *Q. J. G. S.* (1876) xxxii, 408–417. For account of a striking instance of introduction of material giving rise to much tourmaline, see Hawes on the Albany granite, *Amer. Journ. Sci.* (1881), xxi, 21–32.

Besides the brown or blue tourmaline, the metamorphosed rocks consist of quartz, micas, chlorite, andalusite, *etc.* Some of the less altered slates have a spotted character in which the spots are imperfect crystal-grains of andalusite. The more altered rocks are mica-schists.

In rocks metamorphosed by a granitic intrusion metasomatic changes seem to be limited, as a rule, to the elimination of certain volatile constituents, such as water and carbonic acid. When tourmaline has been produced, we have evidence of the introduction of small quantities of other volatile constituents (boric and hydrofluoric acids) from the invading magma. In the neighbourhood of some basic intrusions, however, there seems to have been more important metasomatic change, brought about especially by a transference of soda from the magma to the rocks undergoing metamorphism. Some of the 'adinoles' of the Harz are ascribed to this action. They consist essentially of a fine-textured mosaic of quartz and albite with sometimes other minerals. Mr Teall[1] compares with adinole a rock at Y Gesell near Tremadoc, which has the same mineral composition with the addition of minute scales of mica and chlorite. Several observers have also recorded the existence of isotropic matter in slates near a contact with diabase[2].

Apart from any introduction of soda, *etc.*, a very important feature in the metamorphism of many argillaceous rocks is the abundant new formation of felspars. This is probably a quite common occurrence in the advanced stages of metamorphism, but very careful study is needed to distinguish the felspar from quartz when it occurs in a minutely granular mosaic. Good instances are furnished by the Coniston Flags near the Shap granite.

An example of extreme metamorphism is afforded by the Silurian shales near the New Galloway granite[3]. The rocks consist of quartz, light and dark micas, the former predominating, red garnet, and subordinate felspar. The mica gives

[1] Teall, 219–221.
[2] *Cf.* Hutchings, *G. M.*, 1894, 44, 45, 74, 75.
[3] Miss Gardiner, *Q. J. G. S.* (1890) xlvi, 570–573.

a foliated character to the mass, and the quartz tends to aggregate in little knots or lenticles.

Metamorphism of calcareous rocks. It appears that, under the conditions which rule in ordinary cases of metamorphism by heat, carbonic acid is not driven off from carbonate minerals except in presence of available silica to replace it. Thus a pure limestone or dolomite-rock is not altered in chemical composition by metamorphism. It is, however, at a sufficiently high temperature, recrystallized into a fine or coarse-grained marble, in which all traces of clastic and organic structures are effaced. This is seen locally in the Mountain Limestone against the Whin Sill of Teesdale, in the purer parts of the Coniston Limestone near the Shap granite, *etc.* The Durness dolomitic limestones in the neighbourhood of the large intrusive mass of Loch Borolan pass into thoroughly crystalline rocks, the Ledbeg marbles, some of which are pure aggregates of dolomite[1]. Some of the Glen Tilt limestones again are marbles free from any accessory minerals. In the latter case we seem to have thermal metamorphism on a regional scale.

Many metamorphosed limestones, however, have had sufficient impurities to give rise to various lime-bearing silicates, which are found in the recrystallized limestone as crystals, crystalline aggregates, patches, plumose tufts, *etc.* The chief characteristic minerals have been noted above. Two or more of them often occur in association and sometimes with a regular arrangement. Thus some beds of the Coniston Limestone near the Shap granite enclose large crystals of idocrase in stellate groups or nests, each nest surrounded by a shell composed largely of felspar. The Ledbeg marble, mentioned above, contains salite and its alteration-product serpentine. In the Glen Tilt rocks we find chiefly amphibole-minerals—tremolite, actinolite, and green or even brown hornblende. A band of crystalline limestone near Tarfside in the highly metamorphosed area of Forfarshire has green hornblende, zoisite, felspar, quartz, sphene, and other minerals. The metamorphosed Triassic limestones of Monzoni in the Tyrol are famous for their numerous accessory minerals, idocrase being

[1] Other specimens are of calcite with little or no magnesia.

one of the most prominent. Even a large dyke may produce considerable metamorphism in impure calcareous rocks. Thus the Mountain Limestone near the Plas Newydd dyke on the Menai Straits contains birefringent lime-garnet with polysynthetic twinning and a remarkable lime-bearing analcime. Crystalline limestones with accessory minerals of metamorphic origin may, however, attain a considerable development in areas of 'regional' metamorphism. The cipollino of the Italian geologists is a rock of this kind containing mica and other silicates.

The most striking effects, however, are produced in very impure limestone or dolomite rocks or calcareous shales, slates, or tuffs. In these the carbonic acid is completely eliminated, and the whole converted into a *lime-silicate-rock* (the German 'Kalksilikathornfels' or 'Kalkhornfels'). It appears too that quite a moderate amount of calcareous material in shales, tuffs, *etc.*, suffices to make the metamorphism take this line instead of those described under the head of argillaceous rocks. The metamorphosed rocks consist of aggregates, usually but not always fine-grained and compact, of silicates rich in lime with sometimes quartz, pyrites, or other minerals. Several of these minerals occur in association, giving rise to rocks of complex constitution; and beds differing slightly in the amount and nature of their non-calcareous material result in different mineral-aggregates. Numerous types are illustrated by the metamorphosed Coniston Limestones at Wasdale Head, where they abut on the Shap granite. The purer beds, as already remarked, are converted into crystalline limestones, but the calcareous shales and tuffs have had their carbonate-minerals completely destroyed. The Upper Coniston Limestone is extensively converted into a compact porcellanous-looking rock (hornstone of some authors), in which irregular crystalline patches and grains of pyroxenes and other lime-bearing silicates are recognizable. In some specimens wollastonite predominates, in others augite (omphacite), in others tremolite; and various associations of these and other minerals can be noted in thin slices[1]. Anorthite and probably other felspars are present,

[1] *Q. J. G. S.* (1891) xlvii, Pl. XII, figs. 3, 4.

sometimes in irregular crystal-plates or patches with ophitic habit, sometimes in minute granules. A rather pale brown mica occurs in some beds which have contained a considerable amount of volcanic material. Quartz-grains, probably of clastic origin, have recrystallized as little round patches with mosaic structure. Angular fragments of rhyolite, which are abundant in one bed (an impure calcareous breccia) maintain their sharp outline, while their perlitic cracks, filled by a calcareous infiltration, are now minute veins of brightly polarizing pyroxene. In the compact rocks are sometimes enclosed stellate groups of large crystals (idocrase or augite), each group surrounded by a shell chiefly of plagioclase crystals[1]. A bed in the Lower Coniston Limestone is converted into a mass of garnet and idocrase. The garnet (grossularite) is in good crystals enclosing pyroxene-granules and moulded by the clear idocrase[2]. It shews the optical anomalies noted above. In another bed rounded isotropic garnets are embedded in a finely granular matrix, in which the recognizable elements are apparently wollastonite and augite. Similar types of metamorphism have been described by Lossen in the Silurians near the Ramberg granite of the Harz and by Brögger[3] in the Cambrian and Ordovician strata invaded by the large intrusions of the Christiania district.

The singular rock 'lapis lazuli' has been shewn to be a product of thermal metamorphism in dolomitic limestone. The blue haüyne-mineral lasurite is regarded as an alkali-garnet taking the place of the more usual lime-garnet[4]. This mineral in rude crystal-grains, optically isotropic, forms a large part of the rock; pyrites and a colourless pyroxene (diopside) occur; and unaltered calcite remains in patches of crystalline mosaic.

Metamorphism of igneous rocks. Although the thermal metamorphism of plutonic rocks, lavas, volcanic ashes, *etc.*, has not yet received very much attention, it offers many points of interest and importance. Many of these features

[1] *Q. J. G. S.* (1893) xlix, Pl. XVII, fig. 6.
[2] *Ibid.* (1891) xlvii, Pl. XII, fig. 1.
[3] *Nature* (1882) xxvii, 121 (*Abstr.*).
[4] Brögger and Bäckström, *M. M.* (1891) ix, 243 (*Abstr.*).

are exhibited by the Ordovician volcanic series of the Lake District in the neighbourhood of the granite intrusions of Shap and Eskdale.

The acid igneous rocks are much less susceptible to thermal metamorphism than those of intermediate and basic composition. The rhyolites near the Shap granite do not, as a rule, shew any changes that can be clearly attributed to the effects of heat, and indeed the rhyolite fragments in the calcareous breccia mentioned above preserve, close to the granite, their original structures—cryptocrystalline, microspherulitic, perlitic, *etc.* Where, however, decomposition-products existed in the original rocks, they have given rise to metamorphic minerals. In particular, the green pinitoid substance is converted into a mixture of white and brown micas. The coarsely spheroidal ('nodular') rhyolites illustrate this point. The spheroids had, prior to metamorphism, been altered in the usual fashion into complex nodules having concentric shells of rhyolite substance and of weathering-products. In the metamorphosed nodules the shells of unweathered rhyolite remain unaltered, the flinty siliceous zones are converted into quartz-mosaic with a little mica, and the pinitoid substance is changed into biotite and muscovite. In the cracks which divided the shells there may be a little blue tourmaline.

The fragmental rocks associated with these rhyolites were of much less acid composition, and were probably more weathered prior to the metamorphism. Hence they shew more change, the production of biotite being often observed. As in argillaceous rocks, little spots relatively clear of mica are sometimes present: these shew a crystalline reaction and may be andalusite. The spots disappear with more complete metamorphism, but crystals or grains of andalusite or cyanite are sparingly developed, and finally the rock is completely recrystallized into a finely granular mosaic with a certain amount of biotite, a little opaque iron-ore, *etc.* Relatively large crystals of felspar enclosed in the tuffs are replaced by a new felspar-mosaic, only the general outline of the original crystal being preserved.

In the intermediate and basic rocks metamorphism may give rise to important changes. Diorites are metamorphosed in the Malvern range, the results, however, being complicated by dynamic changes. As described by Dr Callaway[1], the chief effect clearly referable to heat is the replacement of hornblende by a deep brown biotite in the vicinity of an intruded granite[2]. It appears that the hornblende had been, at least to some extent, previously converted into a chloritic mineral. The plagioclase is stated to give rise to white mica. The same author[3] describes the metamorphism of diorite by a granitic intrusion at Galway Bay, where recrystallized plagioclase is observed, and the hornblende has given place to a chloritic mineral, epidote, and rarely biotite.

The Carrock Fell granophyre, in Cumberland, has produced metamorphism in a very basic type of gabbro. In some examples the apatite and iron-ores are unchanged, the turbid felspars become clear, and the augite is converted into green actinolitic hornblende or into biotite. The latter occurs chiefly near the grains of iron-ores, from which it has probably taken up some ferrous oxide and titanic acid[4]. In other specimens the gabbro shews more complex changes.

The metamorphism of diabases by granitic intrusions has been noticed by Allport[5] in Cornwall, by Lossen in the Harz, *etc.* Specimens from these districts shew in various stages the conversion of augite into hornblende and the recrystallization of the felspar. The hornblende produced is mostly green, but in the neighbourhood of the iron-ores (ilmenite) it is sometimes brown. Brown mica or scaly patches of chlorite may be found instead of hornblende, and these often give indications of being formed not directly from augite but from its decomposition-products.

The augite-andesites on the west side of the Shap granite afford fine examples of thermal metamorphism. They had

[1] *Q. J. G. S.* (1889) xlv, 485, *etc.*

[2] On production of a red mica in a diorite, see also McMahon, *Q. J. G. S.* (1894) l, 351.

[3] *L.c.*, p. 495.

[4] *Q. J. G. S.* (1894) l, Pl. xvii, fig. 4. See also Sollas on Carlingford district, *Trans. Roy. Ir. Acad.*, xxx, 493–496, Pl. xxvi, fig. 8, xxvii, figs. 10–16.

[5] *Q. J. G. S.* (1876) xxxii, 407–427. For figs. see Teall, Pl. xvii, and xxi, fig. 2.

undergone considerable weathering prior to the post-Silurian intrusion of the granite. Chloritoid minerals, calcite, chalcedony, and quartz had been formed from the pyroxene and felspar, and were partly disseminated through the rock, but especially collected in little veins and in the vesicles. These weathering-products were the parts most readily affected by the heat. The chloritoid mineral has been converted into biotite, or, where it was associated with calcite, into green hornblende (notably in the vesicles): chalcedonic silica has been transformed into crystalline quartz[1]. The rocks are more altered nearer the granite, and new minerals appear, such as a purplish-brown sphene, magnetite, and pyrites. The plagioclase phenocrysts are replaced by a mosaic of new felspar substance, and finally the whole mass of the rock is found to be reconstituted, the ground becoming a fine-textured mosaic of clear granules.

A more basic type of lava, on the north side of the granite, shews phenomena on the whole very similar to the preceding; but, owing to the larger percentage of lime present, the minerals produced are in part different. Green hornblende predominates over biotite among the coloured constituents of the metamorphosed rocks, and an augite, colourless in slices, is also formed, especially in veins and amygdules. Epidote is another characteristic mineral, and sphene, pyrites, and magnetite occur as before. Especially noteworthy is the formation of numerous lime-bearing silicates from the contents of the vesicles: grossularite occurs, as well as hornblende and actinolite, epidote, augite, and quartz. In the centre of the largest amygdules some residual calcite is found, recrystallized but not decomposed[2].

A basic hypersthene-bearing lava (the Eycott type) is metamorphosed by the Carrock Fell gabbro, the bastite pseudomorphs after hypersthene being converted into a pale hornblende. Here the transformation of the rocks is not always complete, the large labradorite phenocrysts being, as a rule, not recrystallized into a mosaic, but only cleared of their dusty inclusions (fig. 67).

[1] *Q. J. G. S.* (1891) xlvii, 294–298, Pl. XI, figs. 4, 5.
[2] *Q. J. G. S.* (1893) xlix, 360–364, Pl. XVII, figs. 1–4.

The tuffs of basic and intermediate character near the Shap granite have much resemblance to the lavas as regards their metamorphism. Brown mica is the usual ferro-magnesian mineral formed, amphibole being less common. Magnetite is never abundant, and sphene is wanting. The most metamorph-

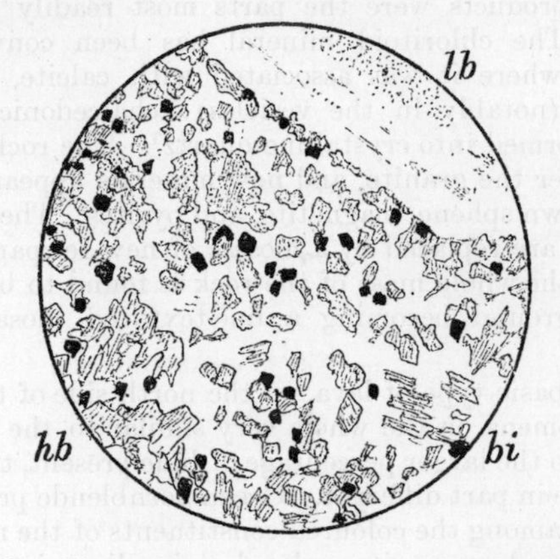

FIG. 67. METAMORPHOSED BASIC LAVA ENCLOSED IN THE GABBRO OF CARROCK FELL, CUMBERLAND; ×20.

The rock was originally a hypersthene-basalt belonging to the Eycott Hill group (see fig. 43). The porphyritic felspars have become clearer (*lb*), their large inclusions disappearing; the pyroxenes or their weathering-products have been converted chiefly into a pale hornblende (*hb*) or locally into biotite (*bi*); the magnetite has recrystallized in good octahedra; and the felspars of the ground-mass are now a clear aggregate, which appears almost homogeneous in natural light [1550].

osed rocks are completely reconstituted into a very fine-textured aggregate of clear granules, in which lie flakes of biotite parallel to either original lamination or cleavage, producing a kind of mica-schist. Felspar crystals enclosed in the tuffs are either transformed into pseudomorphs of epidote or recrystallized into a mosaic[1].

The metamorphism of Carboniferous volcanic tuffs on Dartmoor has been described by Gen. McMahon[2], an interesting

[1] *Q. J. G. S.* (1893) xlix, Pl. XVII, fig. 5.
[2] *Ibid.* (1894) l, pp. 338–366.

feature being the production of the rhombic amphibole anthophyllite in radiating bundles of colourless needles (Sourton Tors, Meldon, *etc.*)

Metamorphism in crystalline schists, etc. On this subject there is not a large amount of information, and it appears that crystalline schists of various kinds are, as a whole, less susceptible to thermal changes than sedimentary rocks. The metamorphism of phyllites and mica-schists has been studied in the Adamello range[1], in New Hampshire[2], on the Hudson River[3], *etc.* In some respects the phenomena resemble those seen in argillaceous strata[4], the production of biotite, andalusite, *etc.*, being characteristic; but there are sometimes quite special peculiarities, in particular the formation of minerals very rich in alumina. Cordierite is sometimes extremely abundant, while pleonaste and other spinels and pure corundum are noted in several localities.

[1] Salomon, *M. M.* (1892) x, 45 (*Abstr.*).
[2] Hawes, *Amer. Journ. Sci.* (1881) xxi, 21–32.
[3] Williams, *Ibid.* (1888) xxxvi, 254–266.
[4] Beck, *M. M.* (1893) x, 265, 266 (*Abstr.*).

CHAPTER XXI.

DYNAMIC METAMORPHISM.

IN this chapter will be noticed some of the effects, mineralogical and structural, produced in rock-masses by the operation of great mechanical forces. Among the mineralogical changes we ought logically to separate those due to pressure from those due to mechanically generated heat, the latter belonging rather to the preceding section. This distinction we shall make so far as our actual knowledge goes.

The consideration of dynamic metamorphism in comparatively yielding rock-masses has already been partly anticipated in the chapter devoted to argillaceous sediments: phenomena more striking, or at least more easily investigated, are now to be noticed in crystalline and other rocks of more stubborn consistency.

Strain-phenomena in crystalline rocks. A frequent effect of strain in the component crystals of a stubborn rock-mass is a modification of the optical properties, which at once becomes apparent between crossed nicols. Instead of being dark throughout for a definite position, a crystal shews dark shadows which move across it as the stage is rotated, owing to the directions of extinction varying from point to point. These *strain-shadows*[1] are best seen in quartz, and are very

[1] Mr Blake styles this appearance 'spectral polarization.' It is spoken of by some foreign writers as 'undulose extinction.'

common in the granitic and gneissic rocks, quartzites, *etc.*, of countries like the Scottish Highlands or the older parts of Norway, which have been the theatre of great crust-movements. Again a mineral such as garnet, normally isotropic, may become birefringent (*e.g.*, in the Eddystone gneiss).

Flexible minerals, such as micas, often shew *bending* of their crystals, or, again, they yield by a shearing movement analogous to lamellar twinning parallel to definite directions known as *gliding-planes* (Ger. Gleitflächen). In some minerals, such as the plagioclase felspars, the gliding-planes coincide

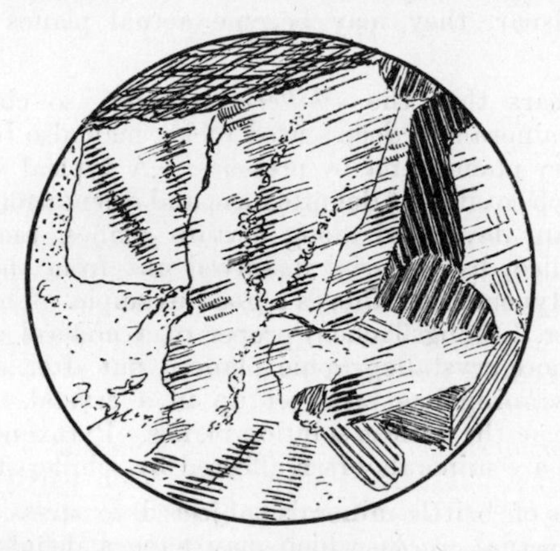

FIG. 68. SECONDARY TWIN-LAMELLATION IN PLAGIOCLASE FELSPAR, DUE TO STRAIN, IN GABBRO, ILGERSHEIM, NAHE DISTRICT; ×20, CROSSED NICOLS.

In places where the strain has been greatest the crystals have yielded along cracks. The mineral at the top of the figure is diallage converted into hornblende [1408].

with natural twin-planes[1], and the *secondary twinning* can be distinguished from original lamellation only by its inconstant character and its relation to bending or other strain-phenomena. It is very clearly seen in such rocks as the norites of Hitterö, Seiland, and Bekkafjord in Norway[2], where the natural twin-

[1] Judd, *Q. J. G. S.* (1885) xli, 363–366; Pl. x, fig. 1.
[2] Rosenbusch-Iddings, Pl. IV, fig. 6.

lamellæ of the felspars are rather broad. Sometimes, in one crystal, the closeness of the secondary lamellæ is seen to increase with the strain, until the crystal has yielded along a crack or a granulated vein[1] (fig. 68). In some rocks there seems to be evidence of the microcline-structure being set up in orthoclase as a result of strain.

Quartz sometimes shews *rows of fluid-pores* marking directions of shearing-strain, and parallel to actual planes of faulting if the crystal has yielded[2]. The lines of pores can be traced through contiguous crystal-grains; or entering another mineral, such as felspar, they may become actual planes of discontinuity.

It appears that the *schiller-structures*[3], so characteristic of certain minerals in deep-seated rocks, may also be produced as secondary phenomena by pressure. A typical structure is that in which cavities of definite form and orientation ('negative crystals') are developed along certain planes, and filled, or partially filled, by material dissolved out from the enclosing crystal. Hypersthene affords a good example. The 'solution-planes' (Ger. Lösungsflächen) proper to a mineral are parallel to one or more crystallographic planes; but after a secondary lamellar twinning has been set up in a crystal, the gliding-planes become the easiest solution-planes. Pyroxenes, felspars, and olivine are minerals often affected by schiller-structures.

Crystals of brittle minerals subjected to stress have often yielded by actual *cracks*, which may have a definite direction throughout the rock, being perpendicular to the maximum tension, and so parallel to the maximum pressure. This is sometimes seen in quartz and felspars, but most commonly in the garnet of granulites, eclogites, gneisses, and crystalline schists (fig. 75). As a further stage, the portions of a fractured crystal may be separated and turned over, or drawn out in the direction of stretching or flowing movement in the solid rock. It is noticeable that quartz shews these phenomena much oftener

[1] See Rosenbusch-Iddings, Pl. IV, fig. 6.
[2] Judd, *M. M.* (1886) vii, 82, Pl. III, fig. 1.
[3] Judd, *Q. J. G. S.* (1885) xli, 374–389, Pl. X–XII; *M. M.* (1886) vii, 81–92, Pl. III.

than felspar: the former mineral, though harder than the latter, is more brittle.

Cataclastic structures. The phenomena of internal fracture and crushing of hard rocks ('cataclastic' structures of Kjerulf) are to be seen in endless variety in some regions of great mechanical disturbance. They may be developed in less or greater degree; they may affect some or all of the mineral constituents of a composite rock; they may or may not tend to a parallel arrangement of the elements. In one type the rock-mass breaks up along definite surfaces of sliding, the material bordering the cracks being often ground down by friction: this is *brecciation in situ.* The irregularly intersecting surfaces divide the rock into angular fragments; but these may be rolled over and their angles rubbed off, so that a 'friction-conglomerate' as well as a 'friction-breccia' may arise, especially along faults and thrust-faults (*e.g.*, Lake District). According as the new structure is on a large or a small scale, the fragments may be recognizable pieces of rocks or portions of constituent crystals of an originally crystalline rock.

Again, we sometimes find the larger elements of a rock—grains of quartz, crystals of felspar, *etc.*—surrounded by a border of finely granular material furnished by the grinding down of the crystal itself and adjacent ones. This is the *morter-structure* (Ger. Mörtelstructur) of Törnebohm. As a further stage, the finely granular portion of the rock may make up the chief part of its bulk, forming a matrix which encloses portions of crystals not yet destroyed but indicating by irregular polarization their strained condition. Beautiful examples are seen among the crushed quartzites and gneisses of Sutherland (fig. 69).

In many cases mechanical forces having a definite direction have caused uncrushed fragments to assume an eye-shaped or *lenticular* form (Ger. Augenstructur) with their long axes perpendicular to the maximum pressure, and so parallel to one another and to any schistose structure in the matrix (fig. 72, *A*). In such cases the crushed matrix usually has a more or less well-marked parallel structure or *schistosity*, in part analogous to slaty cleavage. The final result of the grinding down and rolling out processes is the type of rock named *mylonite* by Professor Lapworth[1], in which, except perhaps for occasional uncrushed 'eyes,' all original structures are lost. In these much crushed rocks the 'eyes' no doubt represent in many cases porphyritic crystals, usually of felspar, in what was once

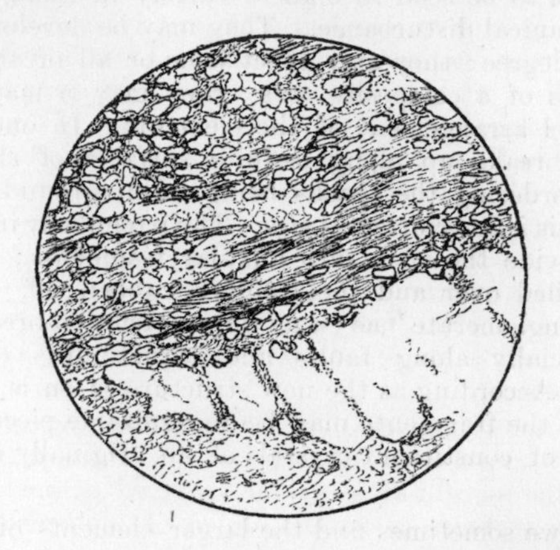

FIG. 69. ADVANCED CATACLASTIC STRUCTURE IN GNEISS, SOUTH SLOPE OF BEINN MOR OF ASSYNT, SUTHERLAND; ×20.

The greater part of the rock is completely broken down and has partly taken on the parallel structure of a mylonite. A large grain of quartz is only partly crushed, and this between crossed nicols shews strain-shadows [1641].

an ordinary igneous rock. It is evident, however, that, in the absence of such indications, it must often be impossible to determine by microscopical study alone the nature of a rock whose original structures have been totally obliterated.

Mineralogical transformations, etc. In extreme stages of crushing of crystalline rocks, the changes produced are by no means purely mechanical. In consequence of the stress and subsequent release a *recrystallization* of minerals may be effected, resulting in the clear, finely granular aggregate

[1] See Page (Lapworth) *Introd. Text-book Geol.*, 12th ed., figs. on p. 107: Geikie, p. 544, fig. 256.

which forms a large part of some dynamo-metamorphic rocks[1]. It must be remembered, however, that thermal metamorphism due to mechanically generated heat may complicate the strictly dynamic changes.

Further, atomic as well as molecular rearrangement has operated in greater or less degree in any dynamo-metamorphic rock not of the simplest constitution. Certain *mineralogical transformations* seem to be characteristic of dynamical metamorphism, being either developed by the action of great pressure or at least facilitated by pressure even when they can also take place without that condition[2]. It should be noticed that in crystalline, and in general in hard, rocks, these mineralogical changes may begin before any important structural modifications are produced. In softer rocks, however, structural changes are the more easily brought about (clay-slates), and new minerals are extensively set up only by more intense action (phyllites).

One characteristic change is the production of colourless mica at the expense of alkali-felspars. The mineral may be formed at the margin of a crystal squeezed against its neighbours or on surfaces of lamination or of movement in a felspathic rock : in such cases it takes the filmy form known as sericite. Or it may replace the interior of a crystal partially or almost wholly. Potash-felspar gives rise to muscovite, soda-felspar to paragonite.

A characteristic alteration in the soda-lime-felspars results in the minutely granular aggregate which has been called 'saussurite,' and is not always of precisely the same nature[3]. The soda-bearing silicate of the felspar separates out as very minute clear crystals of albite, while the lime-bearing silicate, in conjunction with other constituents of the rock, goes to form minerals rich in lime. Zoisite is a characteristic mineral, or its place may be taken by yellow or colourless epidote; and needles of actinolite may also occur. (Compare fig. 70.)

[1] Cf. Teall, p. 175, figures.

[2] See G. H. Williams, *Greenstone Schist Areas of...Michigan*, Bull. 62 *U. S. Geol. Surv.* (1890) Ch. I.

[3] Teall, 149–152. For a somewhat similar process of 'granulation' of plagioclase resulting in a fine mosaic of albite, *etc.*, see Hyland, *G. M.*, 1890, 205–208.

The conversion of plagioclase into scapolite under dynamic action seems to be a more complex process, involving the presence of sodium chloride in solution [1].

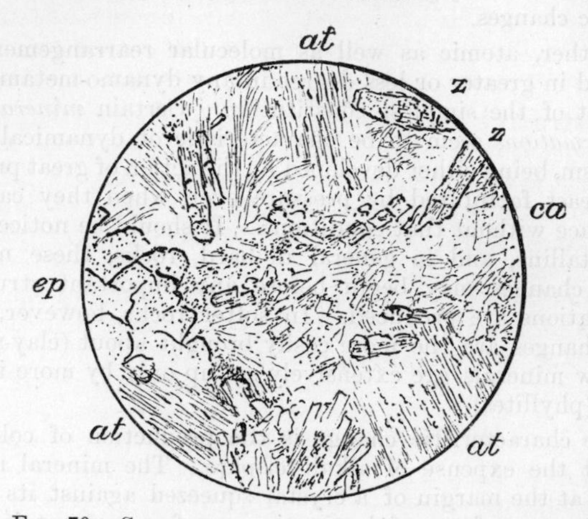

FIG. 70. SAUSSURITE-GABBRO, NORWEGIAN BOULDER ON THE YORKSHIRE COAST; × 20.

The portion figured consists of patches of pale greenish fibrous hornblende or actinolite (*at*), calcite (*ca*), and chlorite, prisms of zoisite (*z*), grains of epidote (*ep*), and little clear crystals of secondary felspar. The so-called 'saussurite' is a similar aggregate on a more minute scale [1049].

Other changes common in dynamic metamorphism are the conversion of olivine into tremolite or anthophyllite and talc, and the production of granular sphene at the expense of ilmenite or other titaniferous minerals. Augite gives rise when crushed to chlorite. The conversion of augite or other pyroxenes into green hornblende is also a common feature in regions of dynamic metamorphism: perhaps this is one of the transformations that should be ascribed to the heat generated in the crushing. It is a very wide-spread phenomenon [2].

[1] Judd, *M. M.* (1889) viii, 186–198, Pl. IX.

[2] See, *e.g.*, R. D. Irving, *Amer. Journ. Sci.* (1883) xxvi, 27–32; G. H. Williams, *ibid.* (1884) xxviii, 259–268; Teall, *Q. J. G. S.* (1885) xli, 133–144.

The borders ('reaction-rims') sometimes noticed at the junction of two different minerals in a crystalline rock have been attributed to dynamic metamorphism. Thus a border of hypersthene may be interposed between olivine and anorthite (norite of Seiland), or a double rim of hornblende and anthophyllite between a lime-felspar and hypersthene or olivine.

Illustrative examples. After the above remarks it will be sufficient to mention a few cases in illustration of what is a very wide and only partly explored field of research. Much valuable information has been published by observers in various European districts, and especially by Lehmann in his work on the Saxon Granulite Mountains, with numerous photographic plates [1]. The most complete study in English of a region of dynamic metamorphism is perhaps that by G. H. Williams of the 'greenstone-schists,' *etc.*, of the Lake Superior region, which further contains a general summary of knowledge on the subject [2]. The dominant types of rocks in the areas there studied have been basic eruptives, probably true lavas in great part, and these are now represented by chlorite- and hornblende-schists. Gabbros, diorites, granites, and quartz-porphyries have also been included, and shew their appropriate types of alteration. The author traces in detail the processes of uralitization, chloritization, epidotization, saussuritization, sericitization, *etc.*, as well as the structural changes undergone by the rocks.

In our own country, and especially in some parts of the Scottish Highlands, the phenomena of dynamic metamorphism are exhibited on an extensive scale [3]. Dykes in the western part of Sutherland shew very clearly the conversion of diabase into hornblende-schist, and an instance of this has been described in detail by Mr Teall [4]. The augite is transformed into green hornblende, and the felspar has recrystallized in

[1] *Entstehung der Altkrystallinischen Schiefergesteine, etc.* Bonn (1884), Atlas.

[2] *The Greenstone Schist Areas of the Menominee and Marquette regions of Michigan*, Bull. 62 of *U. S. Geol. Surv.* (1890) Ch. I, VI, figs. and plates.

[3] See Report in *Q. J. G. S.* (1888) xliv, 429–435.

[4] *Q. J. G. S.* (1885) xli, 133–144, Pl. II; *Brit. Petr.*, Pl. XIX, XX, pp. 197–200.

water-clear grains, while the titaniferous iron-ore has also been altered, giving rise frequently to granular sphene. These mineralogical changes may be produced without any schistose structure, but the massive hornblendic rock further becomes in places a typical hornblende-schist. This is at Scourie: other examples are seen near Unapool, on Loch

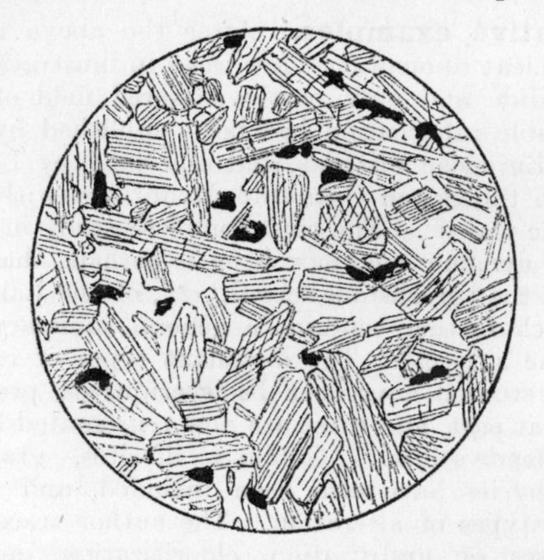

FIG. 71. AMPHIBOLITE OR HORNBLENDE-SCHIST, FROM THE METAMORPHISM OF A DIABASE DYKE, LOCH ASSYNT, SUTHERLAND; ×20.

The rock now consists essentially of idiomorphic hornblende and clear secondary felspar, with some magnetite. The slice is cut parallel to the schistosity, which therefore is not apparent in the figure [1664].

Glencoul, and near Loch Assynt (fig. 71). At Lochinver dykes of enstatite-peridotite pass into an anthophyllite-schist, consisting of matted aggregates of anthophyllite prisms or needles with little patches of brilliantly polarizing talc and large rhombs of carbonates[1].

Near Loch Assynt and in other places the Lewisian gneiss is traversed by zones of crushing, within which the rock is completely reconstituted, and from the granitoid assumes the 'granulitic' structure. The rock so metamorphosed shews a

[1] Mr Teall speaks of one of these rocks as a talc-gedrite-siderite-schist.

rather fine-textured mosaic of clear quartz and felspar, enclosing imperfect crystals of green hornblende and ragged flakes of brown mica instead of the original pyroxene. There is a marked parallel structure and some tendency in the several minerals to collect into little lenticular aggregates. The basic and ultrabasic dykes involved in these crush-zones are metamorphosed in the manner just described.

Some of the above-mentioned changes are perhaps to be ascribed rather to the effects of mechanically generated heat than to pure dynamic metamorphism. In the district farther east there are also some phenomena which seem to point to thermal effects, *e.g.* the production of brown mica in the Torridon Sandstone near the 'Beinn Mor thrust-plane.' But in proportion as the rocks affected give evidence by increasing schistosity of thorough mechanical degradation and sliding movement, those mineralogical transformations which seem to belong to pure dynamic metamorphism become more general. Near the great 'Moine thrust-plane' the sericitization of the acid rocks and the chloritization of the basic ones reach their fullest development in connection with the maximum display of mechanical deformation. Detailed petrographical observations on this interesting district are not yet forthcoming, and the same must be said of the region east of the great thrust-faults, where the complex of gneissic and other crystalline rocks known as the 'Moine schists' is supposed by some to represent the old gneiss and other rocks of the west completely transformed by dynamic agencies.

Illustrations of dynamic metamorphism are furnished in the Central Highlands and in Ireland by various members of the Dalradian series of Sir A. Geikie. The so-called 'green schists' are ascribed by that geologist partly to the crushing of basic lavas and tuffs. Some of these rocks again have the appearance of intrusive diabases, in which every stage of crushing into chloritic schists, *etc.*, can be traced (North Esk, Kincardineshire).

We have seen that under certain conditions the crushing of a diabase may give rise to a hornblende-schist, while under different (perhaps more superficial) conditions a chlorite-schist may result, consisting essentially of chlorite, quartz, and other secondary products, often including calcite or other carbonates. This latter change is often found locally in the neighbourhood of a fault (*e.g.* Portmadoc[1], Ousby Dale[2], *etc.*). Beautiful examples occur among the older dykes of the Isle of Man. Here the carbonates are often dolomite or chalybite. White mica is sometimes produced abundantly, but only in the immediate neighbourhood of planes of slipping. Some of the 'schaalsteins' of German geologists seem to be crushed diabases and dolerites, and many of them contain calcite. Both the hornblende- and the chlorite-mode of alteration are seen in crushed diabases in South Devon and Cornwall, where other dynamic phenomena such as the production of schistose and granulitic structures in granite[3] may also be studied.

Gradual transitions from massive diorite to hornblende-schists may be studied in Anglesey, especially between Holland Arms or Gaerwen and Menai Bridge[4]. In the processes by which these schistose rocks have been produced, the felspar has often been destroyed, and is represented in great part by epidote, which is often abundant. The granular sphene which is often seen is probably derived in part from ilmenite, as well as from the original sphene of the diorite. The hornblende is recrystallized in imperfect elongated crystals of green colour with marked parallel orientation. Locally the place of this mineral is taken by a beautiful pleochroic glaucophane, and a rock near the Anglesey Monument[5] is a glaucophane-epidote-schist, with little trace of any other mineral, except veinlets of clear secondary felspar. The pleochroism of the glaucophane (bright blue to pale lilac) and the epidote (yellowish green to pale yellow) make a slice of this rock a very striking object.

The name amphibolite has often been applied to rocks, usually more or less markedly schistose, in which hornblende is the dominant mineral. Many of them are doubtless the results of dynamic action on diorites and sometimes on diabases

[1] Teall, pp. 216, 217.
[2] *Q. J. G. S.* (1891) xlvii, 524, 525.
[3] Teall, Pl. XLII.
[4] Blake, *Rep. Brit. Assoc. for* 1888, 406.
[5] Blake, *G. M.*, 1888, 125–127; Teall, Pl. XLVII, figs. 1, 2.

and gabbros. Two or three types from the Scottish Highlands have been figured by Mr Teall. An epidote-amphibolite from Glen Lyon, Perthshire[1], consists of green and brown hornblende, epidote, and subordinate quartz, with grains of rutile. A zoisite-amphibolite from near Beinn Hutig, in Sutherland[2], contains pale green hornblende, zoisite, and clear felspar. Such rocks are of special interest as reproducing on a relatively large scale the mineral-associations characteristic of the so-called 'saussurite.'

The 'porphyroids' of some authors are, for the most part, quartz-porphyries more or less modified by dynamic metamorphism. They have received a rough schistosity, which is accentuated by films of 'sericitic' mica. This is a secondary mineral formed at the expense of the felspar, and probably includes both muscovite and paragonite. The rock of Sharpley Tor in Charnwood Forest is a good example. Similar features are shewn by the Llanberis mass of quartz-porphyry at numerous points on its south-eastern edge, especially near Llanllyfni (fig. 72, *B*). Some of the 'porphyroïdes' of the Meuse Valley shew a similar schistose structure with much filmy mica. The same plentiful production of sericite in connection with a secondary schistosity is seen in the acid lavas; *e.g.* the old rhyolites, compact and spherulitic, of the Lenne, in Westphalia.

The phenomena of dynamic metamorphism in argillaceous sediments (phyllites, *etc.*) have received some notice in a former chapter. The other groups of sedimentary rocks have been less studied from this point of view. Some of the effects observable in the arenaceous rocks and quartzites of Sutherland[3], culminating in complete mylonitization, we have already alluded to. Calcareous rocks again are susceptible of considerable transformations, chiefly of the nature of structural rearrangement, when subjected to intense mechanical forces. Excellent examples are afforded by the Ilfracombe and other Devonian limestones, to which Dr Sorby[4] drew attention many years

[1] Teall, Pl. XXVIII, fig. 2.
[2] Teall, Pl. XL, fig. 2. *Cf.* actinolite-schist with zoisite, Pl. XXVIII, fig. 1.
[3] Teall, Pl. XLVI, fig. 2.
[4] *Phil. Mag.* (1856) ser. 4, xi, 26–34; *Presid. Addr.* 1879, *Q. J. G. S.* XXXV (*Proc.*) 57–59. See also Marr, *G. M.*, 1888, 218–221.

ago (fig. 72, *A*). These often shew, not only a highly developed slaty cleavage, but also a deformation of the individual fragments (such as crinoidal remains, *etc.*) of which they are

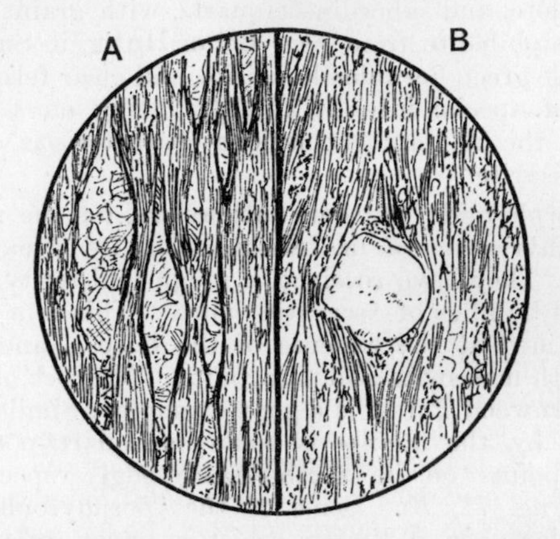

FIG. 72. SCHISTOSE STRUCTURES SET UP BY CRUSHING.

A. Devonian limestone, Ilfracombe; ×20: with uncrushed 'eyes' or lenticles [783].

B. Quartz-porphyry, Llanllyfni, Caernarvonshire; ×20: the schistose structure accentuated by films of secondary mica ('sericite') [87].

largely composed, besides curious phenomena resulting from solution having proceeded at the places of greatest pressure and simultaneous crystallization at the places of greatest relief. The cleaved limestones near Ilfracombe have a microscopic 'eyed' structure, owing to the preservation of uncrushed lenticles of the original rock.

The salite-bearing limestone of Tiree[1], in the Hebrides, also illustrates well the crushing of a crystalline calcareous rock and the production of a fluxional schistose structure of varying perfection. This structure winds past the more resisting grains of salite, felspar, *etc.* (of detrital origin), and in the corners of the 'eyes' so left are uncrushed relics of the original calcite-mosaic.

[1] Bonney, *G. M.*, 1889, 485.

CHAPTER XXII.

VARIOUS CRYSTALLINE ROCKS.

In this final chapter we shall consider briefly certain groups of crystalline rocks, some of very wide distribution, the classificatory position of which is in some doubt, owing to divergence of opinion concerning their origin. It will be evident on consideration that this difficulty arises in great measure from the grouping together under one descriptive name and definition of rocks whose common characteristics have originated in quite different ways. Until more complete knowledge may lead to a true genetic classification, we must be content to bear in mind that such names as 'crystalline schist,' 'gneiss' and 'granulite' do not stand for natural groups, but are of merely descriptive significance; and we notice that various examples of them have already figured in the preceding pages.

Crystalline schists. Under the general title of crystalline schists[1] (Ger. krystallinischen Schiefer) are comprised rocks of distinctly crystalline texture which possess a parallel arrangement of some or all of their elements, often with a tendency to the aggregation of particular constituents into streaks (foliation), and which have in consequence the property of splitting with more or less facility in a definite direction (schistosity).

[1] Many English writers use the name 'schist' simply in this sense. This practice is liable to cause confusion, since the word is used in France (as formerly in this country) for an ordinary shale.

The structures due to the parallel orientation of crystals are various, and should be distinguished. Mr Blake[1] recognizes the 'quincuncial,' in which the crystals that give the structure (*e.g.* flakes of mica) are scattered promiscuously through the rock, but in parallel position; the 'linear,' in which these crystals occur in lines, as well as having a general parallelism; and the 'elemental,' in which the orientation is shewn, not by some particular constituent, but by all the elements of the rock. Further, some degree of aggregation of the several constituent minerals into streaks may give rise to an inconstant banding or to lenticular structures (Ger. 'flaser') on a small scale. The degree of schistosity imparted by these structures depends partly upon the minerals which figure in them, being most marked for flaky and acicular crystals (like mica and actinolite).

It must be observed, as already pointed out, that the meaning thus attached to the term 'crystalline schist' is a purely descriptive one, founded upon structural features which, as we have already seen, may arise in very diverse ways. The rocks included by such a name are not to be regarded as a natural group. A similar remark applies to the special names, mica-schist, hornblende-schist, *etc.*, used for different kinds of crystalline-schists. For another reason, too, such names are lacking in precision, indicating, as they do, only one of the component minerals of a complex rock. Further information may be embodied, if necessary, in epithets (*e.g.* garnetiferous mica-schist) or in compound names (*e.g.* andalusite-mica-schist). Again, such terms as diorite-schist and limestone-schist are sometimes used to indicate that the rock so named has the mineralogical composition of a diorite or a limestone with a schistose structure.

While much difference of opinion exists as to the interpretation of particular areas, it is now generally admitted that the crystalline schists as a whole are metamorphic rocks owing their present distinguishing characters in some cases to thermal, in other cases to dynamic agency[2]. We have studied in the

[1] *Rep. Brit. Assoc. for* 1888, 379, figs. 5–7.

[2] For a summary of views on this question and for much valuable information the student should consult the series of papers on *Les*

two preceding chapters numerous types of crystalline schists (as well as non-schistose rocks) belonging to the two divisions thus indicated. The facts there detailed enable us in a considerable number of cases to tell with some confidence from what type of original rock a given crystalline schist has been produced, and to ascertain whether its metamorphism is the result of heat or of mechanical forces. In other cases one or both of these questions must be left in doubt.

We may remark that, while in rocks resulting from thermal metamorphism foliation and schistosity follow the direction of pre-existing structural planes (laminæ of deposition, cleavage, flow of lavas, *etc.*), in crystalline schists due to dynamic agency the new structures have their direction determined by the forces that produce them, and tend to obliterate, instead of emphasizing, any original structural planes in the mass affected.

Gneisses. The term 'gneiss' is now used to denote, not a rock of some defined composition, but any crystalline rock possessing a *gneissic structure*. By this is to be understood a banded or streaky character due to the association or alternation of different lithological types in one rock-mass or to the occurrence of bands or lenticles specially rich in some particular constituent of the rock. The structure is often found on a relatively coarse scale in rocks of granitoid texture, so that it is to be observed rather in the field or in large specimens than in microscopical preparations. It may, however, be associated with foliation on a smaller scale or with a partial parallel disposition of the elements of the rock. Gneisses, in this sense, may have the chemical and mineralogical composition of acid or intermediate or basic rocks, or may belong to types without parallel among the known products of igneous magmas.

It is generally recognized that gneisses as thus defined have originated in more than one way, but much difference of opinion exists as to the interpretation of the facts in particular districts.

Schistes Cristallins contributed by a number of writers to the International Congress of Geologists at London, 1888; pp. 65–102 of the *Compte Rendu* (1891). The French and German contributions are translated in *Nature*, Sept. 20, 27, Oct. 4 (1888).

We shall note here the three cases which are probably of the most general importance.

(i) We have already seen that gneisses may originate by the thermal metamorphism of some sedimentary (and volcanic) rocks. The New Galloway rocks and the staurolite-, cyanite-, and sillimanite-bearing gneisses of the South-eastern Highlands are examples. The abundance of aluminous silicates is characteristic, and so also is quartz as an essential constituent in rocks with only a low percentage of silica. Under this head are probably to be included such rocks as the biotite-gneiss of the Black Forest, and the rock known as 'kinzigite,' consisting essentially of garnet, biotite, and plagioclase, besides the hornblende-gneisses of the Odenwald, the Wahsatch, *etc.* All these have the chemical composition of sedimentary rocks. Of others, such as the 'red gneiss' of Freiberg in Saxony, Rosenbusch considers that the chemical composition alone is sufficient to remove them from this category and attach them to the true igneous rocks. In this case there remain, as we shall see, two possible explanations of the gneissic structure.

(ii) It appears that gneissic banding may be set up, more particularly in plutonic rocks, by dynamic agency, *i.e.* by the mechanical deformation of a rock-mass originally heterogeneous or of a complex in which one rock was traversed or veined by a different one. In such a case we should expect to find further some degree of foliation and schistosity and usually lenticular structures, quasi-porphyritic 'eyes,' or other characteristic features. Numerous examples have been cited by Reusch from the western coast of Norway and by other observers elsewhere. Mr Teall[1] has applied the hypothesis of mechanical deformation to gabbros, granites, and diorites with gneissic and schistose structures in the Lizard district. Gen. McMahon[2], on the other hand, considers that these structures were impressed on those rocks while still only partially consolidated. He compares the Lizard rocks with the gneissic granites about Dalhousie, *etc.*, in the Himalaya region[3], which he believes to have been intruded in a partially consolidated state and to

[1] *G. M.*, 1886, 481–489; 1887, 484–493.
[2] *G. M.*, 1887, 74–77.
[3] *G. M.*, 1887, 212–220; 1888, 61–65.

have assumed at that time their gneissic and foliated structures.

(iii) There is no doubt that gneissic banding may be an original character in plutonic rocks, dating from the time when the rock in question was still fluid or partly fluid, and due to the different portions of a heterogeneous magma being drawn out in a flowing movement. A remarkable example is described by Sir A. Geikie and Mr Teall[1] in certain Tertiary gabbros in Skye. These rocks shew a striking alternation of light and dark bands due to differences in the relative proportions of the constituent minerals of the gabbro (labradorite, augite, olivine, and titaniferous magnetite). Some narrow bands are composed entirely of pyroxene and magnetite. The authors compare these rocks with the 'Norian' gabbros and anorthosites of North America and, as regards structures, with the Lewisian gneisses of the North-west Highlands.

These latter, apart from the innumerable dykes by which they are traversed, present much variation in character. In the north, between Cape Wrath and Loch Laxford, hornblendic and micaceous gneisses predominate. From Scourie to beyond Lochinver and Loch Assynt the prevalent type is a pyroxenic gneiss[2], consisting essentially of augite or hypersthene (Kylesku), felspars, and quartz. There are also acid types, consisting mainly of felspars and quartz, while, on the other hand, the dominant rock encloses portions very rich in green hornblende. Hornblendic and micaceous gneisses predominate again about Gairloch and Loch Torridon, and a coarse hornblendic gneiss occurs in Lewis (Stornoway) besides other types. Many of these rocks shew in varying degree the effects of dynamic metamorphism, but the authors named consider that much of the banding (as distinguished from foliation) may be ascribed to original conditions attending the intrusion of igneous magmas.

In the South-eastern Highlands (Forfarshire and Kincardineshire) Mr Barrow[3] has described certain micaceous gneisses which are clearly igneous intrusions separable from the metamorphic gneisses, alluded to above, with which they are

[1] *Q. J. G. S.* (1894) l, 645–659.
[2] Teall, Pl. XL, fig. 1.
[3] *G. M.* 1892, 64, 65; *Q. J. G. S.* (1893) xlix, 330–335.

associated. In one phase the rocks consist essentially of quartz, peculiar rounded crystals of oligoclase, muscovite, and biotite. Another phase shews abundant microcline, with a corresponding diminution of oligoclase, while at the same time the white mica predominates increasingly over the brown, and builds larger crystals. The interesting conclusions which the author draws concerning these rocks and their relation to the neighbouring pegmatites should be read in the original.

The gneisses of Ceylon and Southern India, some of which have been described by Lacroix[1], present numerous points of interest. A widely distributed type among the more acid varieties is a microcline-gneiss, very rich in the mineral named and containing orthoclase and quartz, with subordinate oligoclase, biotite, *etc.* The felspars are often crowded with little round or elongated inclusions of quartz ('quartz de corrosion' of French writers) without the regularity of a graphic intergrowth. This is ascribed to secondary corrosion.

The basic gneissic rocks of the same areas are especially pyroxene-gneisses rich in a monoclinic augite which in some varieties has the vivid pink and green pleochroism of hypersthene. In these rocks too are found curious micrographic intergrowths between the ferro-magnesian minerals (pyroxene, hornblende, garnet) on the one hand and felspar and quartz on the other. Lacroix finds scapolite a characteristic constituent of the pyroxene-gneisses here and in other districts.

Granulites (Fr. leptynites[2]). The granulites are fine-textured crystalline rocks consisting of quartz, felspars, and various other minerals, among which garnet is highly characteristic. They shew a remarkable uniformity of grain among the several constituents. There is often a more or less evident parallel orientation of the elements, but no schistosity. Such an even-grained mosaic we have already noticed in some of the products of extreme thermal metamorphism, and again in the rocks resulting from the 'granulitization' of

[1] *Rec. Geol. Surv. India* (1891) xxiv, 157–190.

[2] The 'granulite' of French writers signifies a granite with white and dark micas.

crystalline masses in connection with crushing, while further the same features are found in rocks formed directly from igneous fusion (*e.g.* aplites, like those in the neighbourhood of Dublin). Indeed any petrographical definition of granulite will be found to cover rocks having quite different origins.

It will be sufficient to notice briefly some of the characters of the more or less indefinite group of rocks known as granulites in Saxony and other parts of Europe, where they attain a very considerable development. These rocks have provoked much difference of opinion, but it is now generally believed that many of them are of igneous origin, while they often bear evidence of the operation of mechanical forces either during or after their formation. The varieties of most common occurrence are acid rocks, but there is also a division of basic composition (pyroxene-granulite, or Trapp-granulit of some German writers).

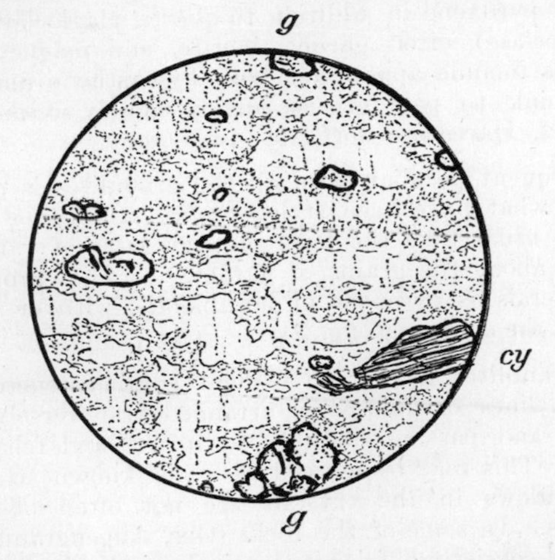

FIG. 73. GARNET-GRANULITE, RÖHRSDORF, NEAR CHEMNITZ, SAXONY; × 20.

Shewing grains of garnet (*g*) and imperfect prisms of cyanite (*cy*) set in a granular aggregate of felspar and quartz. The latter shews a parallel arrangement of its larger elements, and there are rows of fluid-pores traversing the rock at right angles to the parallel-structure [835].

The former contain, in addition to quartz, various alkali-felspars—orthoclase, microcline, and plagioclase, with sometimes microperthite intergrowths. Dark mica is commoner than white as an original constituent, but red garnet is more prominent than either in the usual types of granulites. All these minerals occur in little irregular grains, usually clear except for inclusions of earlier formed constituents. Cyanite, in rude crystals, sillimanite prisms or aggregates (fibrolite), green hornblende, tourmaline, and other minerals may occur, and are taken as marking different types (cyanite-granulite, tourmaline-granulite, *etc.*). Many of these rocks also contain garnet, and garnet-granulite, in which that mineral is the characteristic one, is the most familiar type (Chemnitz district in Saxony, Wartha in Bohemia, Nanniest in Moravia, 'leptynites' of the Vosges, *etc.*).

The pyroxene-granulites are rich in irregular, often rounded, grains of pyroxene in addition to quartz, plagioclase (usually not orthoclase), often garnet, biotite, and magnetite. The pyroxenes include apparently both hypersthene and a pleochroic (pink to pale green) augite closely resembling it (Mohsdorf, Hartmannsdorf, *etc.*, in Saxony).

A frequent peculiarity in all the granulites is the occurrence of what have been styled *centric* structures, of which the most usual take the form of aggregates of various constituents about the grains of garnet, or radial groupings of such minerals as pyroxene or hornblende, with or without a garnet in the centre (*cf.* fig. 74).

In granulites having an evident parallel-structure there are often lines of fluid-pores arranged transversely to that structure and passing through the quartz and felspar alike (fig. 73). This may be noticed as a well-known strain-effect. Strain-shadows in the crystals are not often observed in granulites. In some of the rocks (Ger. Augengranulit) there are 'eyes' consisting of larger lenticular individuals of felspar or quartz-felspar aggregates[1].

[1] Various structures in the Saxon granulites are figured in the Atlas to Lehmann's *Altkrystallinischen Schiefergesteine.*

Of rocks which may be petrographically described as granulites numerous examples are found in the Scottish Highlands. Garnet-granulites are represented, one type consisting of quartz, felspars, garnet, and biotite, with a little muscovite, sphene, and magnetite (*e.g.* Beinn Wyvis). Actinolite-granulites occur, shewing long, imperfect prisms of green actinolite in a clear, even-textured mosaic of untwinned

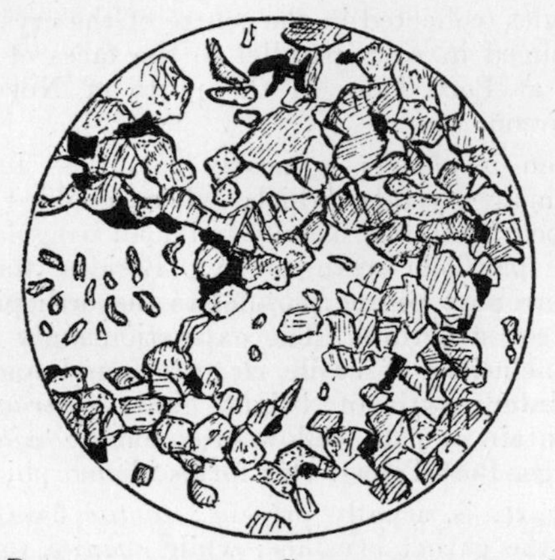

FIG. 74. PYROXENE-GRANULITE, CHEMNITZBACH, NEAR MOHSDORF, SAXONY; ×20.

Much of the pyroxene is hypersthene: the clear portion of the slice is a mosaic of plagioclase felspar and quartz. The rock shews a rude 'centric' structure in the arrangement of the pyroxene-grains [494 *a*].

felspar and quartz, with a little magnetite and small flakes of biotite (Strathan, near Lochinver). Pyroxene-granulites are also found: one type consists of diallagic augite, sometimes hypersthene, and abundant clear plagioclase, with some biotite and magnetite. Garnet is only sparingly present, and there is very little quartz (Badenaban, near Lochinver). Professor Cole[1] has figured a pyroxene-granulite with hypersthene and garnet from near Huntley, Aberdeenshire.

Eclogites. Among rocks of somewhat doubtful affinities must be mentioned the small group of the eclogites. The

[1] *Aids in Practical Geology*, p. 210.

typical eclogite of Haüy consists essentially of an aluminous augite (omphacite) and red garnet, with sometimes quartz, hornblende, actinolite (smaragdite), cyanite, or other accessories. From their mode of occurrence, the rocks are commonly regarded as of true igneous origin.

The dodecahedral or rounded crystals of *garnet* are quite pale in thin slices. They contain various inclusions, such as quartz granules (collected in the centre of the crystal), needles of rutile (ranged in rows parallel to the faces of the dodecahedron, *e.g.* at Port Tana in the north of Norway[1]), little prisms of zircon, *etc.*

The green *omphacite* is nearly colourless in slices. It builds columnar crystals, which, when moulded by quartz, may have good faces, but usually build an irregular aggregate or shew a parallel arrangement. Besides the prismatic cleavage there may be one parallel to the orthopinacoid, or a slight diallagic structure. The extinction-angle rises to 40° or more. Inclusions of rutile, *etc.*, are found, and sometimes a parallel intergrowth of bright green *smaragdite.* Some eclogites contain a pale yellowish green *bronzite* (Bohemia, Fichtelgebirge, Port Tana): this forms idiomorphic crystals.

Clear *quartz* is usually present; *biotite* flakes sometimes cling about the garnet crystals; while *cyanite*, zoisite, zircon, rutile, *etc.*, may be seen in some examples. Iron-ores are not abundant.

The omphacite makes up the bulk of the rock, forming a crystalline aggregate in which the garnet is imbedded, while quartz is always of interstitial occurrence. A clear ring or shell of the last mineral is often interposed between each garnet and the surrounding omphacite. Again, the garnet is sometimes broadly bordered by a 'celyphite'-growth with radial or plumose arrangement and of varying constitution. In a Bohemian example (Chlumiček) it consists of radiating bundles of enstatite prisms: in other cases actinolite, biotite, and other minerals take part in the celyphite-border.

The best known eclogites are from Bavaria (Eppenreuth, with cyanite, *etc.*, Silberbach), the Saxon Granulite Mountains

[1] *G. M.*, 1891, 170, 171.

(Waldheim, with sphene), Carinthia (Saualp, with zoisite), the island of Syra (with glaucophane), and Norway. The only British example of a true eclogite yet described occurs near Loch Duich in Ross [1]. This contains green hornblende, partly surrounding the garnet, and, instead of quartz, a plagioclase felspar occurs in small quantity interstitially or in micrographic intergrowth with the omphacite.

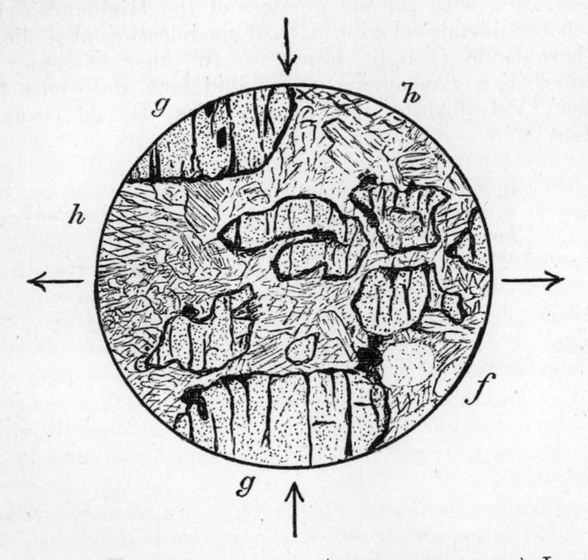

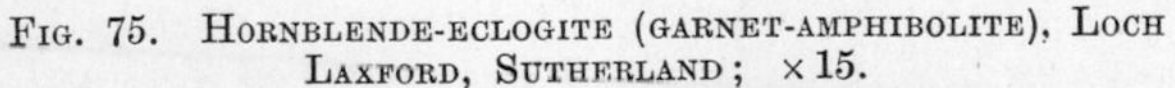

Fig. 75. Hornblende-eclogite (garnet-amphibolite), Loch Laxford, Sutherland; ×15.

Consisting of red garnet (*g*) and green hornblende (*h*), with only a little clear quartz, turbid felspar (*f*), and opaque iron-ore. The arrows shew the directions of the stresses that have operated in the rock, and the brittle garnets are traversed by a strongly marked system of cracks perpendicular to the direction of tension [1254].

Closely allied to the typical eclogites are the rocks styled garnet-amphibolite; in which hornblende more or less completely takes the place of omphacite. Such rocks are found in

[1] Teall, *M. M.* (1891) ix, 217, 218.

Norway, Silesia, and other areas. Prof. Bonney[1] has described one from Beinn Fyn, near Loch Maree, under the name *hornblende-eclogite.* It consists mainly of garnet and green hornblende with some quartz, plagioclase, *etc.* A handsome rock having the same general characters occurs near Loch Laxford, in Sutherland[2] (fig. 75). It will be seen that the only members of the eclogite group yet recognized in Britain are all associated with the old gneisses of the Highlands. In a French hornblende-eclogite, a local modification of a diorite, the hornblende is light brown[3]. In other examples the amphibole is a glaucophane with vivid blue and violet pleochroism (Val d'Aoste, in Pennine Alps, Ile de Groix, in Brittany)[4].

1 *Q. J. G. S.* (1880) xxxvi, 105, 106.
2 *G. M.*, 1891, 171, 172.
3 Fouqué and Michel Lévy, *Min. Micr.*, Pl. VI.
4 Bonney, *M. M.* (1886–7) vii, 1–7, 150–154, Pl. I.

INDEX.

[*Some rock-names are given here which are not admitted as such into the text.*]

CAMBRIDGE: PRINTED BY J. & C. F. CLAY, AT THE UNIVERSITY PRESS.

Pitt Press Mathematical Series.

Arithmetic for Schools. By C. SMITH, M.A., Master of Sidney Sussex College, Cambridge. With or without Answers. Second Edition. 3*s*. 6*d*. Or in two Parts. 2*s*. each.

Key to Smith's Arithmetic. By G. HALE, M.A. 7*s*. 6*d*.

Elementary Algebra. By W. W. ROUSE BALL, M.A., Fellow and Tutor of Trinity College, Cambridge. 4*s*. 6*d*.

Euclid's Elements of Geometry. By H. M. TAYLOR, M.A., Fellow and formerly Tutor of Trinity College, Cambridge.

BOOKS I.—VI. 4*s*. BOOKS I.—IV. 3*s*. BOOKS I. and II. 1*s*. 6*d*.
BOOKS III. and IV. 1*s*. 6*d*. BOOKS V. and VI. 1*s*. 6*d*.
BOOKS XI. and XII. [*In the Press.*

Solutions to the Exercises in Taylor's Euclid. BOOKS I.—IV. By W. W. TAYLOR, M.A. 6*s*.

Elements of Statics and Dynamics. By S. L. LONEY, M.A., late Fellow of Sidney Sussex College, Cambridge. 7*s*. 6*d*.

Or in Two Parts.
PART I. ELEMENTS OF STATICS. 4*s*. 6*d*.
PART II. ELEMENTS OF DYNAMICS. 3*s*. 6*d*.

Solutions to the Examples in the Elements of Statics AND DYNAMICS. By the same Author. 7*s*. 6*d*.

Mechanics and Hydrostatics for Beginners. By the same Author. 4*s*. 6*d*.

An Elementary Treatise on Plane Trigonometry. By E. W. HOBSON, Sc.D., Fellow and Tutor of Christ's College, Cambridge, and C. M. JESSOP, M.A., Fellow of Clare College, Cambridge. 4*s*. 6*d*.

London: C. J. CLAY AND SONS,
CAMBRIDGE UNIVERSITY PRESS WAREHOUSE,
AVE MARIA LANE.
Glasgow: 263, ARGYLE STREET.

Cambridge Natural Science Manuals.

BIOLOGICAL SERIES.

GENERAL EDITOR, A. E. SHIPLEY, M.A.,
Fellow and Tutor of Christ's College.

Now Ready.

Elementary Palæontology—Invertebrate	H. WOODS, B.A., F.G.S.	6*s*.
Elements of Botany	F. DARWIN, M.A., F.R.S.	6*s*.
Practical Physiology of Plants	F. DARWIN, & E. H. ACTON, M.A.	6*s*.
Practical Morbid Anatomy	H. D. ROLLESTON, M.D., F.R.C.P. & A. A. KANTHACK, M.D., M.R.C.P.	6*s*.
Zoogeography	F. E. BEDDARD, M.A., F.R.S.	6*s*.

In Preparation.

Fossil Plants	A. C. SEWARD, M.A., F.G.S.
Text-book of Physical Anthropology ...	A. MACALISTER, M.D., F.R.S.
The Vertebrate Skeleton	S. H. REYNOLDS, M.A.

PHYSICAL SERIES.

GENERAL EDITOR, R. T. GLAZEBROOK, M.A. F.R.S.,
Fellow of Trinity College, Cambridge, Assistant Director of the Cavendish Laboratory.

Now Ready.

Heat and Light	R. T. GLAZEBROOK, M.A., F.R.S.	5*s*.
,, ,, in two separate parts	,, ,,	*each* 3*s*.
Mechanics and Hydrostatics ...	,, ,,	8*s*. 6*d*.
,, ,, in three separate parts		
Part I. **Dynamics**	,, ,,	4*s*.
,, II. **Statics**	,, ,,	3*s*.
,, III. **Hydrostatics**	,, ,,	3*s*.
Solution and Electrolysis	W. C. D. WHETHAM, M.A.	7*s*. 6*d*.

In Preparation.

Electricity and Magnetism	R. T. GLAZEBROOK, M.A., F.R.S.

Now Ready.

Elements of Petrology	A. HARKER, M.A., F.G.S.	7*s*. 6*d*.

Other volumes are in preparation and will be announced shortly.

1/11/95

Press Opinions.

BIOLOGICAL SERIES.

Elementary Palæontology—Invertebrate. By HENRY WOODS, B.A., F.G.S. With Illustrations. Crown 8vo. 6*s*.

Nature. As an introduction to the study of palæontology Mr Woods's book is worthy of high praise.

Saturday Review. The book is clearly and concisely expressed; it conveys much information in a comparatively small compass and cannot fail to be most useful to the student. Not only will it give him clear ideas upon the subject, but with it as a guide he will find his way more easily about the larger works or special memoirs on Palæontology, to the saving of his time and the increase of his knowledge.

Academy. It will be distinctly useful to any student entering on the study of geology.

Science and Art Journal. Geological students will find this admirable work on Invertebrate Palæontology easy reading, and thoroughly up-to-date.... We consider the book a most valuable addition to our scientific literature, and recommend it to all who desire to acquire a sound introduction to a knowledge of the past life-forms of our planet.

Practical Physiology of Plants. By F. DARWIN, M.A., F.R.S., Fellow of Christ's College, Cambridge, and Reader in Botany in the University, and E. H. ACTON, M.A., late Fellow and Lecturer of St John's College, Cambridge. With Illustrations. Crown 8vo. 6*s*.

Nature. A volume of this kind was very much needed, and it is a matter for congratulation that the work has fallen into the most competent hands. There was nothing of the kind in English before, and the book will be of the greatest service to both teachers and students....... The thoroughly practical character of Messrs Darwin and Acton's book seems to us a great merit; every word in it is of direct use to the experimental worker and to him alone.

British Medical Journal. This book will prove a valuable one for the student of practical botany....... The instructions for the study of these and similar facts in the botanical laboratory are set out with great clearness, and the figures illustrating apparatus used and tracings obtained are extremely good, and will greatly help the investigator who avails himself of the guidance of this work.

Glasgow Herald. Mr F. Darwin is well known as an authority on Botany, and the work before us will certainly prove a safe and satisfactory guide to the student in the botanical laboratory.... The directions for work are all clearly given, and for teachers possessing a laboratory a better students' guide to practical work in botany could not be found.

Natural Science. The text throughout is exceedingly clear, and the index full and carefully compiled.

BIOLOGICAL SERIES.

Practical Morbid Anatomy. By H. D. Rolleston, M.D., F.R.C.P., Fellow of St John's College, Cambridge, Assistant Physician and Lecturer on Pathology, St George's Hospital, London, and A. A. Kanthack, M.D., M.R.C.P., Lecturer on Pathology, St Bartholomew's Hospital, London. Crown 8vo. 6*s*.

British Medical Journal. The editor of the "Cambridge Natural Science Manuals" has been fortunate not only in the selection of the above-named subject but also in securing as authors Drs Rolleston and Kanthack...... This manual can in every sense be most highly recommended, and it should supply what has hitherto been a real want.

The Medical Chronicle. "This handbook is an attempt to supply a practical guide to the post-mortem room," say the authors in their introduction, and any competent reader will acknowledge that they have succeeded in their attempt. They have not only supplied the student with a large amount of reliable information, but have done it in a clear and very readable form.

PHYSICAL SERIES.

Heat and Light. An Elementary Text-book, Theoretical and Practical, for Colleges and Schools. By R. T. Glazebrook, M.A., F.R.S., Assistant Director of the Cavendish Laboratory, Fellow of Trinity College, Cambridge. Crown 8vo. 5*s*. The two parts are also published separately.

Heat. 3*s*. **Light.** 3*s*.

Nature. Teachers who require a book on Light, suitable for the Class-room and Laboratory, would do well to adopt Mr Glazebrook's work.

Science and Art. For the practical courses on Heat and Light now forming such a prominent feature in the curriculum of so many of our schools and colleges, these books are admirably suited.

Educational Review. Mr Glazebrook's great practical experience has enabled him to treat the experimental aspect of the subject with unusual power, and it is in this that the great value of the book, as compared with most of the ordinary manuals, consists.

Saturday Review. It is difficult to admire sufficiently the ingenuity and simplicity of many of the experiments without losing sight of the skill and judgment with which they are arranged.

Journal of Education. We have no hesitation in recommending this book to the notice of teachers.

School Guardian. It is no undue praise to say that they are worthy both of their author and of the house by which they are issued.

Teachers' Aid. Text-books of which it would be almost impossible to speak too highly.

Press Opinions.

PHYSICAL SERIES.

Mechanics and Hydrostatics. An Elementary Text-book, Theoretical and Practical, for Colleges and Schools. By R. T. GLAZEBROOK, M.A., F.R.S., Fellow of Trinity College, Cambridge, Assistant Director of the Cavendish Laboratory. With Illustrations. Crown 8vo. 8*s*. 6*d*.

Also in separate parts.

Part I. **Dynamics.** 4*s*. Part II. **Statics.** 3*s*.
Part III. **Hydrostatics.** 3*s*.

Educational Review. In detail it is thoroughly sound and scientific. The work is the work of a teacher and a thinker, who has avoided no difficulty that the student ought to face, and has, at the same time, given him all the assistance that he has a right to expect. We hope, in the interests both of experimental and mathematical science, that the scheme of teaching therein described will be widely followed.

Scotsman. While expounding well the theory of the subject, the book is essentially a practical one for use in large classes in schools and colleges. It is simply and clearly written and has a large number of examples, experiments, and illustrative diagrams; and will be welcome to those who have to instruct beginners in the study of Physics.

Educational Times. We are bound to say that the book is full of good matter, clearly expressed, set out in excellent form and good print.

Educational News. We recommend the book to the attention of all students and teachers of this branch of physical science.

Journal of Education. A very good book, which combines the theoretical and practical treatment of Mechanics very happily.

Machinery. It is quite clear that a great deal of care has been taken in the arrangement of this volume, which will be found of great value to students generally whose initial difficulties have been carefully considered and in many cases entirely overcome.

Knowledge. We cordially commend Mr Glazebrook's volumes to the notice of teachers.

Educational Times. The absurdities which infest books on Mechanics, even the very best, in their language involving the term "force" are absolutely avoided.

Glasgow Herald. The student will also find excellent instructions for the working of experiments in the laboratory.

Technical World. The apparatus used is simple and effective and well adapted for classwork.

London: C. J. CLAY AND SONS,
CAMBRIDGE UNIVERSITY PRESS WAREHOUSE,
AVE MARIA LANE.
AND
H. K. LEWIS, 136, GOWER STREET, W.C.
Medical Publisher and Bookseller.